CAMPUS ECOLOGY

CAMPUS ECOLOGY

A GUIDE TO ASSESSING ENVIRONMENTAL QUALITY AND CREATING STRATEGIES FOR CHANGE

APRIL A. SMITH
and the
Student Environmental Action Coalition

Illustrated by
Thorina Rose

LIVING PLANET
PRESS
LOS ANGELES

Interior design and page layout: Karen Bowers
Cover illustration: Thorina Rose
Cover design: Annmarie Dalton
Editors: Dinah Berland, Eve Kushner

Discounts for bulk orders are available from the publisher. Call (310) 396-0188.

 Printed on Earthcare® recycled paper

ISBN 1-879326-19-1

Manufactured in the United States of America

Library of Congress catalog card number 92-073441

10 9 8 7 6 5 4 3 2 1

DEDICATION

To students around the world who are fighting to protect our natural resources and enhance the quality of human life, and to all the teachers who inspire them.

"Nobody made a greater mistake than he who did nothing because he could only do a little."
—Edmund Burke

CONTENTS

SECTION I: WASTES AND HAZARDS

SECTION II: RESOURCES AND INFRASTRUCTURE

SECTION III: THE BUSINESS OF EDUCATION

SECTION IV: TAKING ACTION

AUTHOR'S ACKNOWLEDGMENTS

There are numerous individuals and organizations who deserve thanks for their support and contribution to this project. Without financial support from Esprit, the Southern California Gas Company, the Mary Reynolds Babcock Foundation, the Jesse Smith Noyes Foundation, Environment Now, the W. Alton Jones Foundation, the Rockefeller Family Fund, and David Zucker, this book would never have become a reality. I would also like to recognize SEAC's fiscal sponsor, the Tides Foundation, with special thanks to Miyoko Oshima.

I am especially grateful to my UCLA *In Our Backyard* team members Jennifer Dill, Gretchen Holmblad, Tamra Brink Mabbott, Anita Glazer Sadun, and Bob Gottlieb for their contributions to the original *Campus Environmental Audit* and this book. I would also like to recognize Owen Byrd and Denis Hayes from Earth Day 1990 for helping to inspire the *Audit* idea and promoting it around the country and abroad. Thanks also to Earth Day Resources for their ongoing efforts to support this project.

I would like to thank the SEAC National Council and staff, with a special thanks to Alec Guettel, Helen Denham, Jeremy Hays, Randy Viscio, Beth Ising, and Adrea Peters. Gratitude is also extended to Nick Keller and Julian Keniry, and all the folks at the National Wildlife Federation's Cool It! Project. And for their help in gathering material, I thank my research assistants Stacy Patterson, Anand Pandian, and Billi Romain. For his legal guidance, SEAC and I thank Spencer Weisbroth.

Numerous individuals have reviewed various parts of the manuscript and offered valuable comments. For their advice and insights, I would like to thank David Orr, Gary Valens, David Eagan, Robin Sherman, Matt Nickademous, Melissa Everett, Susan Englehardt, and David Novak. Thanks also to Dinah Berland for her terrific edits and helpful suggestions. Gratitude is also extended to Cliff Gladstein for his assistance and support.

I would also like to thank my good friend Saul Janson for always "being there" when I most needed support. And I am especially grateful to Josh Horwitz at Living Planet Press for his enduring patience and unyielding faith in this project.

It is becoming increasingly apparent that the our present environmental crisis is evidence of a prior failure of mind and perception—which is to say, a failure of education. The loss of species, topsoil, rainforests, and impending climate change are not primarily technological problems or even economic ones. They are first and foremost problems rooted in how we think about the world we inhabit.

The reasons for this are not difficult to find. "Modern education," according to historian Page Smith, "has its roots in a world view born in the eighteenth century and elaborated, explored, and modified in the nineteenth. " From the vantage point of the eighteenth and nineteenth centuries there were few if any limits to the Earth. Accordingly, academia aimed to extend human mastery over the earth, building on foundations laid down by Galileo, Descartes, and Bacon. But now we know the Earth to be more complex, mysterious, and limited than the creators of modern academia could have realized—and less amenable to management and manipulation than their successors have often assumed. The ecological expertise and ethical wisdom necessary to heal, restore, and preserve the Earth in the coming century will require a different knowledge set than that required to industrialize the Earth in the past century and a half. This recognition underscores the need to rethink how educational institutions operate as well as the content and process of education itself.

Schools, colleges, and universities are beginning to respond to the environmental challenge, more often than not because of student interests and pressures. As a result, many institutions have begun to recycle solid wastes. Some are using recycled paper. A few are vigorously pursuing energy efficiency throughout campus operations. But we have a long way to go towards reducing the environmental impacts of educational institutions to acceptable levels, and towards integrating environmental change into a transformed curriculum.

Campus Ecology uses the campus as a laboratory for the study of resource flows and for the implementation of environmentally sound alternatives. It introduces the kind of analytical abilities and practical skills students will need to address the ecological challenges looming before them. It seeks to solve real problems that are embedded in organizations whose decisions shape our lives and environment. Most importantly, this is a book about educating people to think broadly, observe carefully, and act responsibly.

Underlying *Campus Ecology* is a vision of renewed educational institutions that lead by example, that catalyze change, and that help communities move toward sustainability. I invite you to become an active part of this urgent process of renewal and transformation.

— *David W. Orr*
Chair, Environmental Studies Program
Oberlin College

AUTHOR'S NOTE TO STUDENTS, FACULTY & STAFF

In the mid-1980s, the greenhouse effect and its potential for global warming were only textbook theories. They were incredibly shocking and frightening ideas, but still seemed abstract. After the summer of 1988, however, many of us started to think differently. The scorching heat that raked across the southeastern United States, the agricultural drought, plus reports of medical wastes being washed up on our shores, brought home the reality of how an actual, long-term environmental disaster might feel.

As a college student taking a course in *Introduction to Environmental Studies*, I read textbook treatments of global warming, acid rain, ozone depletion, toxic waste, and overpopulation and felt a strange detachment. There were so many critical problems facing our environment, our very survival was apparently at stake. Yet the connection was seldom made between the pressing environmental dilemmas "out there" and what was being done day-to-day by the people and institutions around us.

A few years later, as graduate students struggling to develop a thesis topic in environmental policy, a group of us in UCLA's Urban Planning program decided to apply our classroom skills and knowledge to solving environmental problems in our own community—the campus. For six months we investigated the university's environmental quality by analyzing documents, evaluating campus practices and decision-making, reviewing regulatory policies, interviewing campus officials, and researching alternatives. Our report, which was titled, *In Our Backyard: Environmental Issues at UCLA, Proposals for Change, and the Institution's Potential as a Model*, was the first study to comprehensively examine the state-of-the-environment of a college campus.

At the outset, some students in the group expressed skepticism about the campus environment's relevance to "the real world." But we soon discovered that our study of UCLA's waste-management practices, energy and water use, air emissions, transportation policies, and purchasing decisions provided direct parallels to regional and global environmental problems. Our campus' practices and policies were both microcosmic reflections of the larger community's actions and also directly connected to them.

In our most vivid experience of this connection, we sorted and measured the campus garbage for a day and thus provided a snapshot of the university's waste composition. This study was very revealing and enabled us to recommend specific, effective source reduction, recycling, and yard-waste composting strategies, as well as provide the administration with information needed to meet its state-wide waste reduction requirements. Studying the campus workplace environment revealed that hazardous substances were not just confined to chemistry and medical labs, but were also endemic to art and architecture studios, the theater department, and photography labs, where the use and disposal of waste materials was less controlled, and therefore, more hazardous.

Finally, we were able to conclude that environmental problems at UCLA mirrored those experienced by institutions and communities throughout the state and the nation. And this knowledge gave us the power to act. By developing a profile of environmental issues on campus, we—as students working together with faculty, staff, and administration—could help create a comprehensive framework for environmental policy and innovation that could influence decision-making at other institutions across the country.

The release of our report sparked an unanticipated wave of inquiries on and off campus, locally as well as nationally. The nationwide interest expressed in the research project spurred us to develop a student guide to studying the campus environment. Thus the *Campus Environmental Audit* was born. As the centerpiece for Earth Day 1990's national student campaign, the Audit became a tool for assessing ecological impacts and implementing change at hundreds of schools in the United States and abroad. One month after the twentieth anniversary of Earth Day, I received a copy of the Audit translated into Japanese by college students from halfway around the world. I couldn't have asked for stronger confirmation that the work begun by "mere students" such as ourselves held far-reaching implications for environmental change.

Since Earth Day 1990, an explosion of environmental activity has occurred at the campus level. Hundreds of student environmental groups have formed on college campuses. In response to student and faculty interest, schools are creating and expanding environmental education programs. College and university presidents are answering the international call to action for environmental literacy. Campus professional associations are placing environmental issues on their national agendas. More than ever before, campuses are recycling, exploring energy-efficiency technologies, purchasing environmentally friendly products, and reducing their use of hazardous substances.

Campus Ecology represents the next generation of the original *Campus Environmental Audit,* updating and expanding its scope. This book is designed to take the environmental issues and principles currently being studied in the classroom and move them outside the classroom doors into the campus community and the larger world. By making environmental knowledge part and parcel of campus environmental practice, students, faculty and administrators have an extraordinary opportunity to act as agents of environmental education and change, starting right here in our own backyards. Moving from theory to practice, we can begin to transform our campuses into living models for an ecologically sustainable society.

— *April A. Smith*
Santa Monica, California

INTRODUCTION: THE GREENING OF THE IVORY TOWER

Your campus is more than just a place to learn, teach, and work. It is also a microcosm of environmental problems facing the larger society, and linked to it ecologically in a myriad of ways. Most college and university campuses contain offices, libraries, research labs, hospitals, residential halls, food services, art studios, theaters, roads, parking lots, recreational and sports facilities, and wilderness areas. And all campuses consume vast amounts of water, food, energy, toxic materials, and other products, and generate a variety of wastes and pollutants. Whether your school is large or small, rural or urban, public or private, it has some impact on the larger environment. With more than 3,000 post-secondary schools in the country, the cumulative effect of these concentrated centers of consumption and disposal is tremendous.

Today's institutions of higher education are not "ivory towers," insulated from the rest of society, but are an integral part of the larger community's physical, social, and economic landscape. As centers for research, teaching, and policy development, colleges and universities possess vast resources and influence. Their economic power, through the products they buy, the investments they make, and the companies they do business with, can create and sustain major markets for environmentally friendly products and technologies.

Faced with urgent and increasing environmental challenges, our educational institutions need to educate and graduate environmental problem-solvers, as well as take responsibility for the ecological impacts of their physical plants. If environmental stewardship is the goal, then auditing the campus environment is an excellent first step toward reaching it. Students, faculty, and staff can share in this responsibility by helping their colleges and universities become laboratories for studying resources flows, environmental hazards, and business practices.

All facets of the campus community are critical to this process. Faculty and students can enrich almost any course by examining some of the environmental aspects of their discipline and adding an environmental project to their syllabi or independent research. They won't have to look far. In addition to directly incorporating information from biology, geology, chemistry, and other natural sciences, environmental studies also makes use of data, technologies, and research strategies common to all the sciences. Teachers and students of political science, sociology, economics, anthropology, history, urban planning, architecture, and other social sciences will also find a wealth of ideas in the environmental studies arena. Engineering students will find it critical to their work. Business majors may wish to examine the environmental impact of various corporate practices, and law students could concentrate on environmental legislation. Students and faculty in art, architecture, theater, and photography, may even eliminate a few health risks by studying the materials with which they work. The list of subjects that can be successfully integrated with environmental studies is virtually limitless—precisely because "the environment" incorporates everything that surrounds us, including the quality of human life.

If the multidisciplinary approach—incorporating a range of environmental ideas into existing courses—is enriching, an interdisciplinary curriculum that focuses directly on environmental issues is even better. In time, environmental studies is certain to become a major course of study at many colleges and universities. If it hasn't happened at your school yet, you can help bring it about.

In addition to the academic sector, a variety of other departments have important environmental effects on campus. Food service personnel, physical plant managers, campus architects, and transportation coordinators are essential to environmental program implementation. They can also teach students about campus operations, budgetary processes, and decision-making. In addition, college and university presidents, trustees, and administrators can provide the leadership for achieving long-term environmental goals. While finding out how the campus works, students can learn effective environmental management principles which will carry over to the workplace. By fostering communication and collaboration between all campus groups, *Campus Ecology* can become a guide toward greater ecological awareness, building bridges between academic and administrative sectors.

Historically, college students have stood at the vanguard of important social movements, whether in the struggle for peace, civil rights, or the international fight against apartheid. During the emergence of the environmental movement, students also played a

HOW TO USE CAMPUS ECOLOGY

This book is designed to provide members of the campus community with a framework, a guide for analyzing the nature and magnitude of environmental issues on campus. Hopefully, these ideas will provide the foundation for planning and implementing recommendations and creating long-term environmental goals for your campus. Each chapter outlines the issue at hand, gets you started with a selection of assessment questions, sources for finding the answers, and then provides a list of additional resources.

The first two sections of the book, "Wastes and Hazards" and "Resources and Infrastructure" outline how to research and analyze campus environmental practices issue by issue. Whether you are studying campus building and design, food policies, or hazardous waste management, *Campus Ecology* outlines a standard set of questions which can be applied to every topic: What's the nature of the problem? What are the types and quantities of materials, products, and resources consumed? How much waste is generated? Where do these materials come from and where do they go? What are the costs associated with the current policies and the alternatives? What are your school's current programs and why were they created? Who is the campus authority responsible for managing that particular issue? What is the regulatory framework and which local, state, and national agencies interact with the university for that area?

The third section of the book, the "Business of Education," addresses the environmental impacts of your institution's research and economic activities and its ethical stance. How our schools teach, what companies they invest in, the type of research they conduct, and whom they do business with all have environmental consequences beyond the immediate borders of the campus.

The fourth section of the book, "Taking Action," provides strategies for working with members of the campus community to create and implement environmentally respon-

critical role as catalysts for change. The phenomenal impact of the original Earth Day 1970 and its twentieth anniversary in 1990 was largely due to student activism. It was a Harvard Law School student, Denis Hayes, who inspired Earth Day 1970, a spark that helped ignite the creation of our current national environmental policy, including the passage of the Clean Air and Water Acts and the formation of the Environmental Protection Agency (EPA). By raising public consciousness about environmental threats, the twentieth anniversary of Earth Day offered citizens around the globe an impetus for working toward environmental changes in their own communities.

sible, sustainable practices. Any meaningful reform process must include the widest possible cross section of the students and community members. For too long, people of color have been excluded from environmental debate and reform despite the fact that minority and low-income neighborhoods are common dumping grounds of industrial polluters. Fighting for environmental justice means addressing the social inequities of the politics of pollution and reaching out to the economically and environmentally disenfranchised.

Although this book focuses on specific environmental impacts, it is important to consider these issues within the broader social and economic contexts in which they exist. Looking at environmental practice through an ecological lens, a range of issues must be taken into account: environmental quality, peace and justice, labor practices, educational access, and local economic development. All these issues are inextricably linked and deserve consideration in your research and advocacy. The planet is our ultimate "common ground," and creating a sustainable future means forging alliances across traditional social and economic boundaries.

The Campus Environmental Audit Response Form in the back of this book will help you document the findings of your research. The form will also allow you to compare information, monitor changes, and evaluate programs from year to year. Additionally, the response form provides a consistent way to collect and track campus environmental information from schools around the country. We strongly encourage you to send a copy of your completed response form to the SEAC national office so that we can maintain an up-to-date database of campus environmental research and advocacy efforts. Additional information, such as news clippings and descriptions of how this book was used, would also be very useful.

Campus Ecology is a beginning: a jumping-off point for an ongoing process of inquiry and innovation. You will undoubtedly come up with excellent ideas of your own. *Campus Ecology* invites you to enter the arena of environmental study and action through your personal area of interest or expertise. If you want to make a difference, here's the place to start.

In the ongoing race to heal our planet and preserve life on Earth for future generations, all participants in campus life must continue to learn, to teach, and to lead. But first, we must learn to communicate. *Campus Ecology* provides a forum for posing questions and discovering answers. This book can also be a link between campuses across the country and around the globe. Write to us about your experiences—your successes and your frustrations. Fill out the "Campus Environmental Audit Response Form" in the back of the book and send us your findings. As our database grows, *Campus Ecology* will become a more effective teaching tool and a more powerful instrument for change.

GETTING STARTED: TAKE A SURVEY OF THE CAMPUS

Your campus represents an extraordinary kind of ecosystem. It is self-sustaining and self-regulating to a large extent. It is also integrally connected to the larger physical, social, and economic environment in which it exists. Before beginning your campus environmental audit, you will need to familiarize yourself with aspects of your school that you may never have considered before. You will need to find out how it works, in general, and how it interacts with the larger community. The following are suggested areas for this important stage of preliminary research.

Your campus library, city hall, chamber of commerce, as well as your campus catalog will serve as your principal resources at this stage of overall information gathering.

Campus Geography and the Physical Plant

◆ **Look at maps of your campus and the surrounding area.** Where is your campus located? Is it in a large city, a small city, or a rural area? Is it a "campus town," serving as the area's principal economic base, or is it only one of many activities in a larger urban center? How much area does your campus cover? How much natural open space and agricultural land does it include? Who are your nearest neighbors? Is your campus set apart from the surrounding community, or are its buildings interspersed with homes and businesses? Does your school operate more than one campus within the larger community? What transportation facilities exist near your campus, and how do the majority of students, faculty, and staff go to, from, and around the campus each day?

◆ **Gather facts on the natural history of your region.** What is the climate of your region: temperature range, average rainfall? Is it subject to drought, flooding, or natural disasters such as earthquakes or hurricanes? What are the geographical features: forested land, rivers, larger bodies of water? What type of wildlife exists in the region? What is the air quality? Do other environmental problems exist, such as polluted waterways, acid rain, threatened wetland areas, depletion of old-growth forests?

◆ **Research the history of your school.** When was it established? What are the ages of its buildings? What are their architectural styles? What is the growth pattern of your school, in terms of population as well as building construction?

◆ **Review the range of facilities located on campus.** How many and what kinds of food services, libraries, classrooms, offices, maintenance facilities, housing, museums, sports facilities, research labs, art studios, trade shops, theaters, parking lots, garages, and off-campus buildings are affiliated with the campus, including its medical centers, if any? Who administrates each type of campus facility?

◆ **Obtain development plans from your school's office of capital programs.** Does your campus have a long-range development plan? When was it adopted and who developed it? Is it being followed and implemented? Is it being updated or revised? Has your school prepared an environmental impact report for a new or projected campus project? These statements can be useful sources of estimated waste generation, traffic, and resource use for new buildings on campus.

Campus Organization and Human Resources

◆ **Contact your school's personnel and admissions offices for campus statistics.** How many faculty, staff, and students does your school now have, both full-time and part-time? Is the campus population growing or stable? Be sure to include estimates of the number of night students and visitors, as well as those present daily. Employee statistics may be available both in absolute numbers and as full-time equivalents (FTE). What percent of students (and employees) live on campus? Do students live nearby or are they dispersed throughout the region?

◆ **Obtain organization charts from the president's office.** How is your university organized? Is it a public or private school? Is it part of a state university system? What is the structure of the governing authority of your school (such as a board of trustees or regents)? What authorities do the state and local governments have over campus operations vs. the school's governing board? What are the educational, business, and professional backgrounds of prominent individuals in the administration? What faculty and staff organizations and unions are represented on campus?

◆ **Contact your student association for a list of student organizations.** How is undergraduate and graduate student government structured? Are there additional student associations representing professional schools and particular student interest groups, such as a public health students association, African-American students association? Does your student government have decision-making power on campus separate from the university administration?

Information Sources for the Campus Environmental Audit

◆ **Research existing and previous campus environmental initiatives.** Was a Campus Environmental Audit completed as part of the Earth Day 1990 national student campaign? Check the Student Environmental Action Coalition (SEAC) national and regional offices or the National Wildlife Federation's Cool It! project to see if they have an audit on file. Have campus environmental impacts been studied as part of a class project or by a campus organization? Contact your environmental studies department faculty and campus environmental groups before you start your work. If an audit has been completed, you can do a "progress report" to evaluate and monitor the changes made on campus since the original research.

WHAT'S THE LAW AND WHO'S IN CHARGE?

In addition to understanding the policy making and governance structure of your college or university, you should familiarize yourself with the array of federal, state, and local environmental laws your school is required to comply with and which public agencies are responsible for their enforcement. While all schools must comply with the same federal laws, state and local regulations are not consistent across the country. Also, if you attend a state university, your school may be exempt from some local environmental and zoning requirements.

The complex web and changing nature of federal environmental statutes, state regulations, and municipal ordinances and codes may at first seem overwhelming. However, sources exist which will help you identify and decipher the information you need. Talking with campus administrators and managers about environmental and occupational health and safety requirements is the best place to start. Also, many schools or college and university associations have public affairs directors or lobbyists in Washington, D.C. or the state capital. These offices may keep track of regulatory requirements and may also be aware of new legislation that may impact your school's environmental practices in the future.

Federal environmental statutes include the Resource Conservation and Recovery Act (RCRA), National Environmental Policy Act (NEPA), Clean Water Act (CWA), Clean Air Act (CAA), Toxic Substances Control Act (TSCA), Federal Insecticide, Fungicide, and Rodenticide Act (FIFRA), Endangered Species Act

(ESA), Safe Drinking Water Act (SDWA), Comprehensive Environmental Response, Compensation, and Liability Act (CERCLA), Emergency Planning and Community Right-to-Know Act (EPCRTKA), Occupational Safety and Health Act (OSHA), and the Pollution Prevention Act.

For information on these federal environmental statutes, review books on environmental law, visit the public affairs department at your campus library, contact the public information office of your regional Environmental Protection Agency (a listing is provided in the Appendix), contact congressional representatives' offices, or consult national environmental organizations, such as the Natural Resources Defense Council, Environmental Defense Fund, or U.S. Public Interest Research Group.

For information on state environmental legislation, contact your state's office of natural resources, environmental affairs, or environmental protection. State departments of health services or occupational health and safety can provide information regarding hazardous substance handling and disposal requirements and other laws relating to the workplace. Some states may also have separate departments, boards, or commissions for air resources, energy, solid waste management, agriculture, parks and wildlife, transportation, and water resources. Contact the office of your state elected officials for a state government directory and a list of legislative committees and their members.

County and regional government agencies have similar departments. These divisions may include transportation and agriculture commissions, air and water quality districts, department of health services, and planning departments. County sanitation and public works departments are also responsible for wastewater treatment and refuse collection.

Jurisdictions and responsibilities are often shared between county and municipal governments. On the local level, your city or town may have a department of environmental affairs, an environmental health department, and a recycling office. Also check with the board of public works, the mayor's office, and the offices of city council members. Your local fire department should be very familiar with regulations regarding hazardous substances.

HOW TO USE THE FREEDOM OF INFORMATION ACT

The Freedom of Information Act (FOIA) is a tool you can use to get access to materials that may be denied to you by government sources. Passed by Congress in 1966 and strengthened in 1974, the FOIA is a federal law providing public access to government records and information. With a few exceptions, the law permits individuals to obtain information from any federal agency. Many states have similar laws (check with your state's Attorney General's Office). Following is an outline of the basic steps you need to take when using the Act:

Determine what you want. Before writing the letter, determine which agency has the information you are seeking. Be as specific as possible. Find out the exact department and address. Should you be writing to a local headquarters or a Washington, D.C. office? Determine specific dates or time periods, locations, and names of individuals who were involved with creating, receiving, or storing information, if possible.

Write a letter. State that your request is made pursuant to the Freedom of Information Act. Ask for the records you are looking for. State that you wish to be

◆ **Trace the origins and destinations of campus resource flows.** Where do campus water, energy, materials, and food come from? Where does garbage and wastewater go? To the extent possible, take field trips to local utilities and power plants, farms, forests, feedlots, manufacturing plants, material recovery facilities, recycling centers, sewage treatment facilities, water sources and distribution facilities, landfills, and incinerators. Tracing your school's waste products and natural resource use "cradle-to-grave" will provide a valuable portrait of how your campus is interconnected with the larger ecosystem.

◆ **Read campus and local community newspapers.** What effects do your school and its facilities have on the local community? Does the school provide cultural events or other services for the community? Is it the major employer in town? Are the community and the university at odds over growth or other issues? Does your university influence events or policies at city, county, and state levels? Has your school newspaper reviewed any campus environmental issues in recent years?

◆ **Review the course catalog.** What classes require the use of chemicals or other hazardous materials? These may include chemistry, art, photography, biology, agriculture, and

contacted before incurring fees exceeding a specified limit. (You are entitled to as much as 100 pages of free copying and two hours of free search time.) You can request that any fees be waived since the information is for educational purposes. Write that you would appreciate a response within ten working days, as prescribed by law. Sign the letter, print your name, and provide an address and phone number. Keep a copy for your records.

Appeal, if necessary. If you don't receive a reply within a reasonable period of time, follow up with a phone call. Send a copy of your letter marked "second request." When you do receive a reply, check to see if it is adequate. If it is not, you may appeal to the agency for more information if you file within thirty days of receipt of the response. If the agency states that they are denying access to information for a particular reason, they are responsible for proving that the records are being rightfully withheld.

Freedom of Information Act

Adler, Allan, and Ann Profozich. *Using the Freedom of Information Act: A Step by Step Guide*, 1983. Washington, D.C.: Center for National Security Studies.

The Handbook on How to Use the Freedom of Information Act: Holding the Government Accountable to Its Actions. Church of Scientology, 1989; to order write: the Director of Public Affairs, 1404 N. Catalina Street, Los Angeles, CA 90027; (213) 661-0836.

Step-By-Step Guide to Using the Freedom of Information Act, ACLU Foundation, 1992; 22 Maryland Avenue N.E., Washington, D.C. 20002; (202) 544-1681.

various trade classes. Be sure to include extension and night courses offered to the general community. What environmental courses are offered? Can you identify classes and/or professors that may help provide ideas for campus environmental programs?

◆ **Contact local environmental and community groups.** What are the specific regional environmental problems to which your campus contributes? These groups can be valuable resources both in your research and for advocacy.

◆ **Obtain a copy of your school's Administrative Policy and Procedures Manual.** This manual should provide information about existing environmental policies and departmental obligations on campus.

◆ **Check the "Resources" in this book, and use the library.** What more do you want to find out? If you need additional background information on certain topics before you begin your research, use the Resources list at the end of each chapter to find the books and make the contacts you need. Also use your library to find current journals and use any databases they may have.

Research Tips

Once you have familiarized yourself with the structure and operations of your school and you have clearly defined the scope and specific focus of your research project, you can then start to develop a system for collecting data and information. The following tips will help expedite the research process:

◆ **Know what information you are seeking and where you can find it.** Each of the following chapters provides you with assessment questions and research sources. Identify the specific individual(s) who will be providing you with the data and information you need. Inform them about your project and let them know that you will be sending them a list of questions. If you will be reviewing records and documents yourself, set up an appointment with those departments.

◆ **Write up your research questions.** Write up the questions you need answered for each individual who will be providing you with information. Be as specific as possible. Make sure to request information early in your research to allow plenty of time to get through the red tape! You may wish to use the Campus Environmental Audit Response Form questions provided in the back of the book or modify them to your purposes. Inform your sources that, if possible, you would like to receive the information between ten days to two weeks from the time they receive your questions.

◆ **Follow up with a phone call.** After one week, call your sources to make sure that they received the questionnaire and to clarify any questions they may have.

◆ **Use the Freedom of Information Act.** If you are denied access to government documents during any part of the research process, you can use the Freedom of Information Act (FOIA) as a tool to request the information that you are looking for. Guidelines for how to use the Act are provided in this section.

◆ **Review and evaluate your data.** When you receive the information you are looking for, review it to make sure it's complete and that you understand what has been provided. If you need assistance interpreting data, faculty members, original sources, or staff from outside environmental organizations can assist you.

◆ **Thank your sources.** If campus administrators or other individuals have taken a considerable amount of time to gather information for you, show your appreciation for their efforts by writing a thank-you letter. You may wish to send them a final copy of your report, an executive summary, or press clippings when the project is completed.

◆ **Develop a tracking system.** In a file, keep a list of all your information sources, their addresses, phone and fax numbers, the dates of your conversations, and other correspondence, including the dates that you sent and received information. All of this information will be important for documentation in your report and will help you maintain efficiency throughout the research process.

SECTION I:
WASTES AND HAZARDS

YOU MAY THINK OF COLLEGES AND

universities as merely institutions of higher learning that produce educated graduates. They are also intensive human habitats that generate a wide array of waste materials, ranging from solid waste to air emissions to sewage. Some hazardous

SECTION 1: WASTES AND HAZARDS

substances on campus—such as radioactive materials and pesticides—may be so well-hidden, you might not even know they exist. In this section we'll track the flow of wastes and hazards on campus, then explore innovative approaches to reducing and preventing waste generation at its source.

Each year, Americans throw away about 180 million tons of trash. The United States generates twice as much waste, per person, as does any other country, and the amount of garbage we produce is on the rise. The Environmental Protection Agency estimates that 80 percent of existing landfills with permits will close within twenty years. The proliferation of waste can largely be attributed to the growing use of packaging materials, "throwaway" convenience items, and other disposable products. Campus communities generate huge amounts of waste, much of which could be recycled, reused, or composted. In addition, the use of chlorine-based bleaching agents in the production of paper products results in the production of highly toxic substances that eventually enter the waste stream. Simply by posing some of the following assessment questions to your campus managers, you can help sensitize them to how much more they may be able to save—in terms of cost as well as resources. Before you begin, make sure you have a clear understanding of the following terms and concepts: solid waste, landfill, incineration, recycling, composting, source reduction, food recovery, waste stream.

Assessment Questions

◆ How much solid waste does your campus generate a year? Figures should be given by volume (cubic feet) or by weight (pounds or tons). Have any waste composition studies been conducted? What are the components of campus waste? What percentage of community solid-waste-generation does your school account for? (See "Research Sources:" A,B,C,G,E)

◆ Who is in charge of solid waste disposal contracts? Do different entities on campus (such as fraternities, sororities, residence halls, medical center) have separate contracts with waste haulers, or is the entire campus covered under one contract? What is the principal waste-hauling company for your campus? (B)

◆ What are the total costs of disposal per year, the cost per ton, and the disposal fee structure? By what percentage have those costs increased in recent years? (B,C)

◆ Where does campus garbage go? How much of it is landfilled, incinerated, recycled, and composted? (B,C,E,F,G)

◆ Are there landfills or incinerators on campus or in the community? Do these disposal sites pose a health hazard to residents? What are the environmental records of these facilities? (C,F)

◆ Does your campus have a recycling program? When was it started? What materials are recycled? Is the program run by students or by the university administration? How is the program funded? Are recyclables separated from waste at the source (such as offices or residence halls) or after it is collected? Does the municipality in which your campus is located operate a recycling program? Is it voluntary or mandatory? (B,C,E,F,G,H)

◆ What percentage of the total waste stream is yard waste? (Usually this figure is between 15 and 30 percent.) Are landscape clippings mixed with or kept separate from other campus wastes? Does your campus use landscape clippings as compost or mulch? (A,D)

◆ How do the various food vendors on campus dispose of food waste? Have any food recovery or food compost programs been initiated? (A,E)

◆ What programs exist on campus to promote source reduction and reuse in order to reduce the quantity of waste generated? (B,H)

Research Sources

Ⓐ Your own waste-stream analysis may be your best resource here (see "How to Conduct a Campus Solid Waste Stream Composition Study," page 6). Use it to find out what kinds of wastes your campus generates.

Ⓑ The facilities department may be contacted for information about garbage volumes, costs, collection process, and disposal contracts. Specific contract arrangements may be the responsibility of the purchasing office.

Ⓒ Representatives from the waste-hauling company may be contacted for additional information about waste costs, quantities, and collection procedures. Custodial staff are also valuable sources of information.

Ⓓ The facilities department can tell you who is responsible for landscape maintenance. Ask maintenance managers about yard waste disposal and find out if they have a composting program.

Ⓔ Food service managers can provide information about disposal of food waste and the use of plastic and paper disposable serviceware in campus cafeterias. Check state laws regarding food-donation policies.

Ⓕ Your city or town public works department, local recycling centers, community environmental groups, and your state's solid-waste-management board can all provide information about how their waste-management policies affect your campus. State and local offices may also have survey information on campus recycling programs.

Ⓖ Your campus newspaper(s) should be able to tell you how much newsprint is distributed on campus and what they do to encourage recycling.

Ⓗ Individual campus departments may have their own waste-reduction programs; contact them for information.

EDUCATION ENVIRONMENT ENLIGHTENMENT

CAMPUS PROFILE:
University of Colorado at Boulder

Pioneer Recyclers

One of the first and best campus recycling programs in the country is C.U. Recycling, which was begun in 1976 at the University of Colorado at Boulder. The program—which is overseen by the University of Colorado Student Union and staffed by a recycling services director, students, and community service volunteers—collects separated recyclables from every campus building. Over 60 percent of the students and staff regularly participate in the effort.

In 1989 alone, 625 tons of newspaper, glass, aluminum, office paper, cardboard, phone books, batteries, and plastics were collected, as well as yard waste; 300 gallons of used motor oil; 40 cubic yards of used clothing, books, and appliances; and 25 boxes of used textbooks to be given to libraries of developing nations. To reduce the amount of waste generation, the university is training its staff to use electronic-mail, encourages double-sided copying, as well as the use of recycled paper products, reusable mugs, retreaded tires, and washable dishes. In all, 22 to 25 percent of the total waste stream was diverted. The program is supported by an extensive public education campaign, which includes press releases, public service announcements, newspaper articles, and audio-visual materials, as well as orientation for incoming freshmen. Future plans include a household-hazardous-waste-reduction program, alternative chemical-waste disposal, and increased recycled-product procurement.

HOW TO CONDUCT A SOLID-WASTE-COMPOSITION STUDY

Conducting a group waste-composition study will provide an excellent snapshot of what's in your school's waste stream, and will form a basis for designing a waste-management plan that promotes reduction, reuse, and recycling. This is primary research in campus anthropology. It is also a great way to learn first-hand about the kinds of wastes your campus generates—the first step to doing something positive about it. The timing of the study is important. Try to do the analysis during a time that reflects the average level of campus activity (mid-semester or quarter, and mid-week). Remember that the time of year will also affect the results of your research. For example, more yard waste will probably be generated in summer and fall than in winter. Team up with a group of students, faculty, and staff—possibly dividing into smaller groups to assess different locations on campus—and then get ready to dig in, step-by-step, as follows:

❶ Materials: Gather the following materials and resources before you begin your garbage audit: sorting tables, a large scale for weighing the waste, separate bins for each sorting category, gloves, surgical masks, a calculator, and materials for recording data. Arrange for your campus newspaper to cover the event—it's a real photo opportunity!

❷ Safety: Wear protective clothing such as long-sleeved shirts, pants, gloves, and surgical masks. Conduct the waste-composition study under faculty or staff supervision. Avoid selecting garbage from areas which may contain hazardous materials, such as medical and chemistry facilities, unless you have professional guidance.

❸ Select Campus Areas: Select three to six areas on campus that represent distinct waste-generation locations, such as residence halls, food services, administration, student union, academic buildings (separate physical sciences and liberal arts, if possible). Designate a sorting site where you can set up tables and conduct your analysis.

❹ Do a Trial Waste Analysis: Prior to the actual waste-composition study, you will find it helpful to conduct a preliminary analysis, using a small sample of garbage (five bags, for example). This will help determine the appropriate waste categories and will improve your methodology.

❺ Collect Garbage: Find out from waste-collection services what their schedule is for the locations under study. Then randomly collect at least five bags of garbage from dumpsters at each campus location prior to daily waste pickup. Label the bags according to their collection points.

6 Calculate Weight and Volume: Once you've transferred all of the garbage to your sorting site, calculate the total weight and volume collected from each region before you begin sorting. Remember to weigh the sorting containers.

7 Sort the Waste into Categories: At a minimum, sort your waste into two broad categories: "recyclable" and "other." A more detailed breakdown of recyclable waste may help you to distinguish between those materials which may be recycled for reprocessing now and those which may be marketable in the future. For example, the technology exists to recycle tin-plated steel cans, phone books, and lower grades of paper. However, markets may not exist for these materials in your area. In addition, food waste can be composted, but it may take time to establish a composting program. Use the following list as a guideline for choosing categories for your composition study:

Recyclable products: white paper, colored paper, computer paper, newsprint, corrugated cardboard, glass, steel cans, aluminum, PET and HDPE plastics*, yard waste

Other materials: other plastics, other paper, other metals, food waste, hazardous substances

* High Density Polyethylene—used to make milk jugs and juice containers
Polyethylene terephthalate—used to make large soft drink bottles

Carefully sort each bag of garbage into categories. Once you've completed the sorting for one region, weigh your containers of material (subtracting the weight of the container itself) and record the figures. The waste volume for each container can be calculated by measuring the depth of the waste in the container multiplied by the cross-sectional area (length times width) of the container. Or, if you know the capacity of a given container (e.g., a ten-gallon bucket), use it to measure total volume.

8 Using the Information: If you don't know the total amount of waste that a particular area generates over a given period of time, present your figures as a percentage. You can say, for example, that newspaper represents about 15 percent of the waste generated from the food service area on campus. If you do know the total weight of all food service wastes, you can multiply that percentage by the total weight to estimate the total amount of waste in each category.

Because you can only analyze a small amount of the total campus waste stream in a single day, your figures will have limited accuracy and should be used conservatively. They will, however, provide important information about the general types and quantities of waste your school produces.

Recommendations

A successful waste-management policy supports a resource-conserving hierarchy: source reduction and reuse first, composting and recycling next, incineration and landfilling last. Purchasing products made from recycled materials is critical to "close the loop." As you participate in discussions to help develop better waste-management policies, you may wish to include a combination of the following ideas:

✔ Finding new ways to promote source reduction and reuse. *In food services:* Encourage the sale of reusable mugs, allowing a discount on beverages for customers who bring their own mugs. Use permanent dishes and utensils or reusable plastic when possible. *In offices:* Reuse corrugated cardboard, interdepartmental envelopes, and other office supplies. Establish photocopying guidelines that encourage the use of half-sheets and double-sided copies. *In student stores:* Train employees to ask customers, "Do you need a bag?"

✔ Establishing a campus-wide recycling program. The program should be supported and managed by the administration and should include a practical means of separating a variety of materials (white and colored bond paper, computer paper, glass, aluminum cans, various types of plastics, corrugated cardboard). The program must target students, staff, and faculty.

✔ Promoting organic composting and mulching. Yard wastes and some kitchen wastes can be composted and used as fertilizer on campus or sold to markets off-campus. Woody yard wastes can be simply chipped for use as a mulching ground cover.

✔ Establishing food recovery programs. Unused food from school cafeterias can be donated to local community centers and homeless shelters, providing food for those in need and diverting tons of food waste from landfills. Another alternative is "pulping" food (grinding and drying it) for sale to local farmers.

✔ Starting "buy recycled" programs and encouraging the purchase of nontoxic paper products. Policies and programs that create a market for recycled and nontoxic products are critical to the success of recycling efforts and more responsible manufacturing methods (see "Procurement Policies," page 59).

✔ Purchasing office equipment that helps reduce waste, such as copy machines with double sided capability and fax machines that use recyclable bond paper.

RESOURCES

Books, Publications, Articles

ASUCLA Recycling Program. *How to Start a Recycling Program on Your Campus,* March 1991. 308 Westwood Plaza, Los Angeles, CA 90024-1640; (310) 206-7589.

Biocycle: Journal of Waste Recycling. JG Press, Inc., Box 351, 18 South 7th St., Emmaus, PA 18049; (215) 967-4135.

Blumberg, Louis, and Robert Gottlieb. *War on Waste: Can America Win Its Battle With Garbage?* Washington, D.C.: Island Press, 1989.

Campus Recycling Workshop Packet. Conference held at the University of Southern California, May 3, 1991. Packet contains articles, contacts, resources and information on how to start a campus waste-reduction program. Local Government Commission, 909 12th Street, Suite 205, Sacramento, CA 95814; (916) 448-1198.

City of Los Angeles, Integrated Solid Waste Management Office. *53 Simple Things Universities and Colleges Can Do to Reduce Waste: Case Studies of University Source Reduction, Recycling, and Composting,* May 1991. Board of Public Works, City Hall, Room 365, 200 N. Spring Street, Los Angeles, CA 90012; (213) 237-1444.

Environmental Defense Fund. *Coming Full Circle: Successful Recycling Today,* 1988. 1616 P St., N.W., Washington, D.C. 20036; (202) 387-3500.

Facing America's Trash: What Next for Municipal Solid Waste? 1989. Washington, D.C.: Congress of the United States Office of Technological Assessment.

Leadership America. *The Wastoid Handbook: Recycling and Waste Management: A Guide for Campus Leaders,* 1989. 1600 Two Turtle Creek Village, Dallas, TX 75219; (214) 526-2953.

Minnich, J. *The Rodale Guide to Composting.* Emmaus, Penn.: Rodale Press, 1979.

Pollock, Cynthia. "Mining Urban Wastes," *Worldwatch* Paper 76, 1987. 1776 Massachusetts Ave., N.W., Washington, D.C. 20036.

Resource Recycling: North America's Recycling Journal. Resource Recycling, Inc., P.O. Box 1054, 1206 N.W. 21st Ave., Portland, OR 97210; 1-800-227-1424.

Solid Waste Management: Planning Issues & Opportunities, 1990. Chicago: American Planning Association, Planning Advisory Service Reports.

The Sourcebook on Solid Waste Management at the University of Michigan, 1992. Plant Grounds and Waste Management Department, 226 E. Hoover, Ann Arbor, MI 48109-1002; (313) 764-3400.

Waste Not; The Weekly Reporter for Rational Resource Management. Work on Waste USA, Inc., 82 Judson, Canton, NY 13617; (315) 379-9200.

Institutions & Organizations

Association of College Unions-International, 400 East Seventh St., Bloomington, IN 47405; (812) 332-8017.

Center for the Biology of Natural Systems, Queens College, CUNY, Flushing, NY 11367; (718) 670-4180.

Cool It! National Wildlife Federation, 1400 16th Street, N.W., Washington, D.C. 20036; (202) 797-5435. Request issue packets for disposables, recycling, and composting.

Environmental Action Foundation, Solid Waste Alternatives Project (SWAP), 1525 New Hampshire Ave., N.W., Washington, D.C. 20036; (202) 745-4879.

Environmental Defense Fund (EDF) Recycling, 257 Park Avenue South, New York, NY 10010; 1-800-CALL-EDF.

Gildea Resource Center, 930 Miramonte Drive, Santa Barbara, CA 93109; (805) 963-0583.

National Association of College and University Food Services (NACUFS), 1405 South Harrison Road, Suite 303-304, Manly Miles Building, Michigan State University, East Lansing Blvd., Lansing, MI 48824; (517) 332-2494.

National Recycling Coalition (NRC), 1718 M St., N.W. Suite 294, Washington, D.C. 20036; (202) 659-6883.

Rutgers Recycling Center, Rutgers University, Dudley Road, New Brunswick, NJ 08903; (908) 932-5858.

Solid Waste Information Clearinghouse Hotline (SWICH), P.O. Box 7219 Silver Spring, MD 20910; 1-800-677-9424.

The University of Colorado Recycling Program, UMC 331, Campus Box 207, Boulder, CO 80309; (303) 492-8307. The office is an excellent clearinghouse for campus recycling information.

U.S. Environmental Protection Agency, Office of Solid Waste, Waterside Mall, Room 2802, 401 M Street, S.W., Washington, D.C. 20460; (202) 382-4675.

Educational institutions are not generally thought of as hazardous-waste generators, and in fact they typically produce much less hazardous waste than industrial facilities. But many activities on campus do use hazardous substances, and they generate a wide variety of wastes. Laboratory chemicals create the largest category of university hazardous wastes, but toxic substances are also used in art, architecture, photography and theater departments as well as in maintenance work and university research activities. While most chemicals used in science departments are strictly regulated and the atmosphere highly controlled, the use of potentially harmful arts-and-crafts materials and maintenance products are often poorly monitored for proper handling and disposal procedures. Hazardous substances found on college campuses can also pose a significant threat, not only to the natural environment but also to human health. (For a detailed review of the potential environmental health hazards found on college campuses, see "The Workplace Environment," page 36.)

Before beginning your assessment of hazardous wastes, make sure you know what types of substances you are looking for and what the best ways are to dispose of each. Also, you may wish to find out more about microscale lab techniques.

Assessment Questions

◆ Which departments and research facilities on campus use hazardous substances and generate chemical wastes? What types of chemicals are used? What types of wastes are generated? Remember to include such departments and activities such as art, architecture, theater, printing, and photography in your research. (See "Research Sources:" A)

◆ How much hazardous waste does your campus generate? How has this figure changed over the past five years? (A,C,D)

CAMPUS PROFILE:
Bowdoin College, Brunswick, Maine

Microscale is Better

In the late 1970s, several chemistry professors at Bowdoin College began looking into ways to improve laboratory air quality in their aging chemistry building. Finding that upgrading the ventilation system would be extremely costly, they considered modifying the laboratory program instead. They discovered that while standard techniques for measuring and identifying chemicals have improved greatly, standard chemistry textbooks still called for using chemicals in

◆ Are waste collection, handling, labeling, and waste-disposal procedures clearly documented? Where and how are hazardous wastes stored? How are hazardous art materials, such as paints and turpentine, discarded? (A,B)

◆ Where does campus hazardous waste go? How much waste is disposed of by various means: recycling, incineration, landfilling, etc.? What are the environmental records of these hazardous waste facilities? (A,C)

◆ Has your school ever been cited for violation of local or state hazardous-waste regulations? What health and safety problems have been associated with the use, transportation, and storage of toxic substances and hazardous wastes at your school? (A,B,C,D,E)

◆ What are the financial costs associated with hazardous-waste disposal? How have these costs changed over the past five years? (A)

◆ Which waste disposal company collects hazardous materials from campus? Does this company pack the wastes or is that done by university staff? (A)

◆ Are there clear records tracing wastes from their point of origin to their ultimate disposal? (A)

quantities nearly as large as were used at the turn of the century. Thus, they began developing miniaturized versions of standard laboratory experiments, using quantities a hundred to a thousand times smaller than before.

A test section of Bowdoin's introductory organic chemistry course began using the revised experiments in the spring of 1981. After a few years of trying out the new techniques, professors found that these "microscale" methods had far more advantages than just improving laboratory air quality. Overall lab safety was improved, students generally achieved better results than in standard lab sections, and they could finish experiments more quickly. Furthermore, the cost of purchasing chemicals and disposing of wastes was greatly reduced.

In 1985, the Bowdoin professors published articles describing their new techniques, and soon thereafter wrote an organic chemistry textbook describing many of the microscale experiments. They also began publishing a newsletter about microscale, called "Smaller is Better." Since that time, several other microscale texts have appeared, numerous glassware manufacturers have begun producing specialized glassware for use in microscale experiments, and more than 400 U.S. colleges and universities now have microscale organic chemistry lab programs, leading to a tremendous cumulative improvement in lab safety and reduction in hazardous-waste generation.

◆ What is being done on your campus to minimize the quantity of hazardous substances used and waste generated? Have microscale chemistry techniques or surplus chemical exchange programs been initiated? What results have these efforts produced? (A,B)

Research Sources

◈ The campus environmental health and safety office, or its equivalent, is your most important source of information about the use and disposal of hazardous substances. This office should be able to give you information about the quantity and type of hazardous wastes they collect, as well as about campus waste-handling procedures.

Ⓑ Professors, lab technicians, and service personnel in all departments that handle hazardous substances, such as biology, zoology, chemistry, art, photography, architecture, and theater art, may be contacted. Ask about their procedures for waste disposal.

Ⓒ The staff in charge of waste disposal should be able to provide you with their records. Because university hazardous wastes are so diverse, a single figure for the total quantity may not be available. However, a good first step is simply to ask them for whatever summary figures they may have. Look through hazardous-waste manifests (the papers filed for each shipment of hazardous waste, describing what is being shipped). Add up the figures by type and by location.

D Local and state environmental and public health agencies and the EPA office in your region may be contacted for hazardous-waste generator reports filed by the university, or for records of university violations. You have the right to obtain such information from public agencies, but you may have to submit a written request and/or make an appointment to look at the files in their offices.

E The Resource Conservation and Recovery act (RCRA) requires "cradle-to-grave" tracking of hazardous wastes. In addition, "right-to-know" laws at both the federal and state levels require users of hazardous materials to submit information about materials used and about accidental releases of hazardous substances to regulatory agencies.

Recommendations

Typically, laboratory wastes are packed in lab packs—large drums containing several small containers of wastes in an absorbent packing material—then landfilled or incinerated. Some wastes, such as organic solvents and waste oil, can be collected in bulk and either recycled or burned as supplemental fuel. Technology is also available for recovering silver salts from photochemicals. Here's an opportunity to make your campus a safer as well as a more ecologically responsible place. Meet with the appropriate people to discuss the following:

✔ Implementing an ongoing program to educate students and staff about proper handling of hazardous substances and proper waste-disposal procedures.

✔ Making sure there is a clear and comprehensive system for tracking all hazardous wastes generated on campus and filing the required RCRA manifests.

✔ Redesigning the laboratory section of chemistry courses to reduce the amount of chemicals used. Where possible, use microscale techniques to reduce chemical amounts by a factor of a hundred or thousand.

✔ Establishing a surplus chemical exchange program to collect hazardous materials from researchers that no longer need them and make them available to other users so that they are not discarded. Educate researchers to buy only as much of a chemical as they need, rather than buying in bulk and ending up with a lot of waste.

✅ Improving practices to ensure proper disposal and waste reduction of toxic art supplies, such as paints, thinners, and photographic chemicals—for example, using settling tanks to recapture turpentine.

RESOURCES

Books, Publications, Articles

Ashbrook, Peter C., and Peter A. Reinhardt. "Hazardous Wastes in Academia," *Environmental Science and Technology,* 1985, 19, (12) 1150–54.

California Department of Health Services, Alternative Technology Section. *Waste Audit Study—Research and Educational Institutions.* Los Angeles: Ralph Stone and Company, August 1988.

Gordon, Ben, and Peter Montague. *A Citizen's Toxic Waste Audit Manual,* 1990. GreenPeace USA, Toxics Campaign, 1436 U Street N.W., Washington, D.C. 20009; (202) 462-1177.

Management of Hazardous Wastes From Educational Institutions: A Report To Congress, U.S. EPA, Office of Solid Waste, April 1989, EPA/530-SW-89-040 A.

Mayo, D. W., R. M. Pike, and S. S. Butcher. *Microscale Organic Laboratory,* 2nd ed., New York: Wiley, 1989.

Mills, J. L., and M. D. Hampton. *Microscale Laboratory Manual for General Chemistry.* New York: Random House, 1988.

National Research Council, *Prudent Practices for Disposal of Chemicals from Laboratories,* Washington, D.C.: National Academy Press, 1983.

Rachel's Hazardous Waste News. Weekly information letter of the Environmental Research Foundation (ERF), P.O. Box 73700, Washington, D.C. 20056; (202) 328-1119.

Sanders, Howard J. "Hazardous Wastes in Academic Labs," *Chemical and Engineering News* 1986, 64 (5), 21–31.

Smaller is Better. Newsletter of Microscale Organic Chemistry Programs, Bowdoin College Department of Chemistry, Brunswick, ME 04011.

U.S. EPA. *Guides to Pollution Prevention: Research and Educational Institutions.* Cincinnati: Risk Reduction Engineering Laboratory, Center for Environmental Research Information, Office of Research and Development, 1990.

U.S. EPA. *Management of Hazardous Wastes From Educational Institutions: A Report To Congress,* Office of Solid Waste, April 1989, EPA/530-SW-89-040 A.

Williamson, K. L. *Microscale Organic Experiments.* Lexington, Mass.: Heath, 1988.

Institutions & Organizations

Citizens Clearinghouse for Hazardous Wastes, Inc. P.O. Box 6806, Falls Church, VA 22040; (703) 237-CCHW.

Department of Chemistry, Bowdoin College, Brunswick, ME 04011; (203) 725-3000.

Environmental Action Foundation, 1526 New Hampshire Avenue, N.W., Washington, D.C. 20036; (202) 745-4870.

GreenPeace USA, Toxics Campaign, 1017 W. Jackson Blvd., Chicago, IL 60607; (312) 666-3305.

National Toxics Campaign, 1168 Commonwealth Ave., Boston, MA 02134; (617) 232-0327.

RCRA/Superfund Hotline, 1-800-424-9346. In Washington, D.C., call (202) 382-3000.

TOXNET, a computerized toxic release inventory database, National Library of Medicine, Specialized Information Services Division, 8600 Rockville Pike, Bethesda, MD 20894; (302) 496-6193; 1-800-638-8480. Available at many university libraries.

U.S. Environmental Protection Agency, Office of Toxic Substances, East Tower of Waterside Mall, Room 521, 401 M Street, S.W., Washington, D.C. 20460; (202) 382-3763.

U.S. Public Interest Research Group, 215 Pennsylvania Ave., S.E., Washington, D.C. 20003; (202) 546-9707.

Low-level radioactive wastes may be generated in research or medical activities on your campus. Since radioactivity cannot be destroyed, disposal of radioactive materials, even the low-level wastes produced on campuses, can present serious problems. Most radioactive materials and wastes are regulated by the federal government. States are required to comply with handling procedures and guidelines set at the federal level and contract with federal agencies for permission to license and regulate radioactive materials. Your school can improve radioactive waste-management procedures by enforcing rigorous safety training and monitoring programs, by encouraging the use of less hazardous materials in research, and by employing state-of-the-art disposal methods.

Before beginning your assessment, make sure you understand and can discuss biodegradable scintillation fluids; various solvents, such as toluene and benzene; the difference between radioactive aqueous and solid wastes; radioactive source terms; radioisotopes; half-life; "rad waste" incineration; and the official policy concerning radioactive wastes, known as ALARA (as low as reasonably achievable).

Assessment Questions

◆ What departments and activities on your campus (biomedical research, medical therapies, energy, and scientific research) use radioactive substances? Who has access to radioactive materials? (See "Research Sources:" A,B)

◆ What kinds of research use radioactive materials? What are the applications of this research? Is this research commercial and/or academic? (A,B,C)

◆ Are researchers who are involved with radioactive materials encouraged to use safer, alternative scintillation fluids instead of solvents such as toluene and benzene, which are classified as hazardous substances? What incentives are offered to encourage the use of such alternatives? (A)

◆ What are the quantities of radioactive substances used and wastes generated on campus annually? How has this figure changed over the past five years? (A,B,F)

◆ What sorts of radioactive materials are used in campus facilities? Which isotopes? What are their half-lives? What kind of radiation do these isotopes emit? (A,B,C)

◆ What are the terms and conditions of the university's license for handling radioactive substances? Who are the persons named on the license and responsible for campus safety? (A)

CAMPUS PROFILE:
University of Wisconsin, Madison

Being Radioactively Safe

As the second-largest nonmilitary licensee of radioactive materials in the country, the University of Wisconsin has developed a model radioactive waste-management program. Its Office of Radiation Safety focuses on reducing the amount of hazardous substances used in research and on-campus radioactive waste incineration and encourages rigorous training, monitoring, and safety in the disposal of these dangerous waste materials.

◆ What is the licensing agency and what is the regulatory agency (these may not be the same)? Who is licensed (usually the institution)? (A)

◆ Is there a campus office in charge of radiation safety? Is there a safety handbook? What are their safety guidelines? Is the required official policy regarding radioactive materials (ALARA) well-understood and followed by those who use radioactive materials on your campus? (A)

◆ What kinds of training programs exist for radiation safety? (Training is required by the Nuclear Regulatory Commission.) (A,D)

◆ What agency inspects your campus? What agency is responsible for enforcement? Have there been any violations of license terms? (A,D,E)

◆ How is radioactive waste disposed of (wastes might be landfilled or incinerated)? Where is the disposal site? Does your school have an on-site rad waste incinerator? Are any decayable radioactive wastes stored on campus? (A)

◆ What are the annual costs and who pays for the disposal of radioactive wastes? How have these costs changed over the past five years? (A,G)

◆ Does your campus have a nuclear reactor for research purposes? (A)

In 1987 the University of Wisconsin became the first radioactive research licensee to convert to using biodegradable scintillation fluids instead of hazardous solvents, such as toluene and benzene. The Safety Office purchased a vial crusher, which separates liquid from glass and plastic vials. These alternative liquids can be sewered by law, thereby eliminating a hazardous waste stream and associated disposal costs. Switching to safer fluids in the scintillation process has reduced the university's disposal expenses by $45,000 a year. Campus researchers were intimately involved in the process of identifying the best solvent-free scintillation materials and have a strong incentive to use them as well. If substances such as toluene are used when alternatives are available, then the department's research grant is required to cover the disposal costs.

Careful separation of radioactive aqueous and solid wastes has also resulted in significant volume reduction. One approach has been to control the "source term" of the waste, for example, by replacing disposable paper-work-area covers with washable stainless-steel trays. Additionally, all ash from radioactive wastes that are incinerated on-campus is closely monitored and tested to ensure proper and safe landfill disposal or shipped to a radioactive waste-disposal facility. Incinerated decayable radioactive wastes are stored at a campus site until they can be safely landfilled. While the university once generated 20,000 cubic feet of radioactive solid waste each year, the Safety Office now ships only about 70 cubic feet, although the number of radioactive material orders has remained constant.

Research Sources

◆ The campus office responsible for radioactive-materials purchasing, monitoring, and waste disposal (usually the radiation safety office or environmental health and safety office) will probably be your first contact. Interview the campus radiation safety officer. Request information regarding the amount and type of waste generated, disposal procedures, costs, and safety practices.

◆ Medical center administrators and professors in charge of radiation safety and disposal are excellent resources for their areas. They can explain the uses of radioactive materials in the medical field.

◆ Researchers who work with radioactive substances and science department administrators responsible for safety or disposal practices may also be willing to share their procedures. Also interview lab managers or students who work with low-level radioactive materials.

◆ The federal Nuclear Regulatory Commission, Department of Transportation, and Environmental Protection Agency all regulate some aspect of low-level radioactive materials and have general information on handling and disposal.

E State governing agencies vary; jurisdiction may be within a health services department or environmental monitoring agency. This agency can give you specific information on campus licenses and inspections.

F A local community group may be interested in radioactive materials and waste. They may be aware of specific activities and practices on campus and any past problems or controversies.

G The radioactive-waste contractor for your school can provide information about where wastes are discarded.

Recommendations

If you choose to become involved with the radioactive waste safety on your campus, you will need to meet with administrators and faculty at various levels. When meeting with the safety committee, your discussions may focus on the following issues:

✔ Establishing a comprehensive safety program dealing with use, handling, and disposal of radioactive materials. The facility's license requires certain safety procedures. The safety committee's task will be to ensure that these measures are being carried out. Stricter measures may be appropriate under certain circumstances.

✔ Developing a campus policy to reduce radioactive waste generation. One simple waste-reduction measure is to use smaller vials for scintillation fluids or to identify nonhazardous scintillation fluids.

✔ Conducting a study to compile all nonradioactive methodologies that could be used to accomplish the same results as those now using radioactive materials.

RESOURCES

Institutions & Organizations

Englehardt & Associates, Inc. Radiation Consultants. 2800 S. Fish Hatchery Road, Madison, WI 53711-5399; (608) 274-4227.

Nuclear Information & Resource Service, Campus Radiation Survey Project, 1424 16th St., N.W., Suite 601, Washington, D.C. 20036; (202) 328-0002. Request fact sheet called "Radiation Issues on Campus."

Nuclear Regulatory Commission, Woodmont Building, 8120 Woodmont Avenue, Bethesda, MD 20814; (202) 492-4661.

Physicians for Social Responsibility, 1000 16th St., Washington, D.C. 20006; (202) 785-3777.

Union of Concerned Scientists, 26 Church Street, Cambridge, MA 02238; (617) 547-5552.

University of Wisconsin, Madison, Safety Department, 317 N. Randall Ave, Madison, WI 53715; (608) 262-8769.

An estimated 3.2 million tons of medical waste are generated each year in the United States. While the proliferation of disposable medical supply products has tremendously improved sanitation and convenience in the medical field, it has also dramatically increased the quantities of medical waste generated by hospitals, health clinics, and research facilities. Medical waste includes the following:

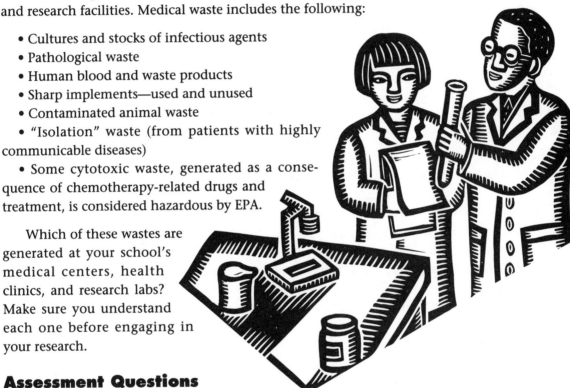

- Cultures and stocks of infectious agents
- Pathological waste
- Human blood and waste products
- Sharp implements—used and unused
- Contaminated animal waste
- "Isolation" waste (from patients with highly communicable diseases)
- Some cytotoxic waste, generated as a consequence of chemotherapy-related drugs and treatment, is considered hazardous by EPA.

Which of these wastes are generated at your school's medical centers, health clinics, and research labs? Make sure you understand each one before engaging in your research.

Assessment Questions

◆ Where is medical waste produced on campus? How much is generated? How has this figure changed over the past five years? (See "Research Sources:" A,B)

◆ How is this waste disposed of (incineration, landfill, etc.)? Where is the disposal site? Is it bagged separately and labeled appropriately? (Standard industry practice is to use red bags to indicate infectious and other medical wastes.) Has your school violated any disposal regulations regarding medical wastes (e.g., ocean dumping)? (A,B,D)

◆ What are the school's annual medical-waste disposal costs? How have these costs changed over the past five years? (A)

◆ Is medical waste tracked and manifested? (A)

◆ What are your state's regulations for handling and disposal of medical waste? (A,C)

◆ Approximately what proportion of the waste is plastic? (A,B)

CAMPUS PROFILE:
Thomas Jefferson Medical School
Philadelphia, Pennsylvania

On Second Thought

Employees of the Thomas Jefferson Medical School Hospital recently took a look at the hospital's entire waste stream and found that much of the waste that the hospital classified as "infectious" really wasn't. In one lab, for example, they found that lab-generated computer paper was being put in the red bags reserved for infectious waste, instead of being recycled. They took measures to correct such problems, and their infectious waste generation decreased by about 25 percent from levels three years before. The hospital is also making a concerted effort to minimize its use of disposable plastic instruments, and has gone back to using some reusables. The entire medical school campus is following in the footsteps of its hospital.

◆ If medical waste is incinerated, are plastics included? (Incineration of polyvinyl chloride [PVC] plastic can produce toxic air emissions.) (A,B,D)

◆ How could the amount of medical waste be reduced? (A,B)

◆ About what proportion of this waste is noninfectious, but mixed with infectious waste? Could it be separated? (A,B)

◆ Where are hospital linens cleaned? Laundry workers may be exposed to infectious materials. Are appropriate safety measures practiced? (A,B)

◆ What other safety precautions are taken to reduce exposure to medical waste? (A,B)

Research Sources

◆ Hospital, lab, and health center administrators dispose of waste themselves or supervise the process. A specific staff person should be in charge of medical-waste disposal.

◆ Other personnel who directly handle medical wastes include nurses, orderlies, housekeeping staff, waste collectors, and disposal-firm personnel.

◆ State legislators may also be of help. Ask your state senators and representatives if they or anyone else in the legislature has introduced a bill regarding medical wastes. This legislation may be a useful source of background information.

◆ Your school's medical-waste contractor can tell you where campus waste is discarded.

Recommendations

Your role in bringing together medical center administrators and others can help produce a campus-wide public-health initiative. Here are some ideas to include:

✔ Improving the way medical waste is handled and tracked.

✔ Reducing the use of disposable medical materials, and providing for proper handling of reusables.

✔ Reducing the incineration of plastic, especially PVC waste.

✔ Separating noninfectious waste from other medical wastes and recycling, if possible.

RESOURCES

Books, Publications, Articles

"There's More than Healed Bodies Coming Out of that Hospital." *Environmental Action.* July/August 1988.

U.S. Congress Office of Technology Assessment. *Medical Waste Management Background Paper.* October 1988. Available from Superintendent of Documents, Washington, D.C. 20402-9325; GPO Stock Number 052-003-011380-7.

U.S. EPA. *Guides to Pollution Prevention: Selected Hospital Waste Streams.* Cincinnati: Risk Reduction Engineering Laboratory, Center for Environmental Research Information, Office of Research and Development, 1990. EPA # 625/7-90/009.

Institutions & Organizations

EPA RCRA Hotline, 1-800-424-9346. Ask for a copy of the "Standards for Tracking and Management of Medical Waste."

CHAPTER 5: WASTEWATER AND STORM RUNOFF

The treatment and disposal of wastewater and storm runoff are topics of great concern in many rural, agricultural, and urban areas because of the need to protect beaches, rivers, and waterways from pollution. Wastewater, or sewage, generated by your campus probably flows to a treatment plant that provides primary (removes solids), secondary (adds bacteria), or tertiary (filters and chlorinates) treatment. The treatment plants in many urban areas are reaching capacity so that excess sewage flows into bodies of water only partially treated. Storm runoff comes from water collected in storm drains from streets and parking lots. In some areas, storm runoff is collected and treated with sewage; in others it flows directly to a body of water. The amount of wastewater or runoff from a site, such as a campus, is rarely measured as such. Instead, estimates of wastewater generation are usually based on water use.

Wastewater and storm runoff may contain hazardous materials such as paint, oil, grease, and heavy metals that can disrupt bacterial treatment and contaminate receiving waters. Through proper design and maintenance, however, much of our wastewater can be reused for a variety of purposes, such as landscape irrigation. Due to heightened concerns about water conservation, many communities are now turning their gray water into a resource to help reduce water consumption.

Because water-disposal problems are often hard to track down, you will need to know what to look for before beginning your research. Start by taking a field trip to your local waste-treatment or reclamation facility. Make sure you understand the following concepts: gray water, groundwater, wastewater, and reclaimed water.

Assessment Questions

◆ How much wastewater (sewage) does your campus produce annually? How is the amount estimated, or is it metered? Amounts may be estimated per person or per square foot of building space, or based on how much water is used. (See "Research Sources:" A,B,H)

◆ Where is your wastewater treated? Is the treatment plant at or near capacity? What percentage of plant capacity does your campus generation account for? (A,C,E,F)

◆ What type of treatment does wastewater receive? Where is the product discharged? (C,F)

CAMPUS PROFILE:
Pennsylvania State University
University Park

Watching the Water

During each day of the school year, Penn State's innovative water reclamation policy prevents 3.5 to 4 million gallons of wastewater from being discharged into local streams, which feed into Chesapeake Bay. An on-campus sewage treatment plant provides wastewater treatment for the campus and some surrounding areas. The treated and chlorinated water is then piped three or four miles and used to irrigate university agricultural sites to provide hay, oats, and corn for the School of Agriculture's livestock and state gameland areas. The ground itself serves as tertiary treatment for the water, and eventually it rejoins underground water tables. Since 1983, virtually all of the wastewater from the university has been reused for irrigation, instead of draining into waterways.

◆ Does your sewage treatment facility have a water reclamation program? Does your school use any reclaimed water from the facility? How much? Does your school reuse any of its gray water? (A,D)

◆ Is campus wastewater monitored for hazardous contaminants? How often and by what agency? Have contaminants been found? (A,B,F)

◆ Are there education programs to encourage people not to pour chemicals down the drain? (A,H)

◆ Is storm runoff collected and treated with sewage or is it discharged directly into a body of water? Which bodies of water are at risk? (C)

◆ Are vehicles washed on streets? Is water used to clean streets, parking lots, and sidewalks? What precautions are taken to prevent washing oil, dirt, and debris into storm drains? (A,B)

◆ Does your campus have on-site farms, orchards, or feed lots? If so, are pesticides and fertilizers used? How are animal wastes disposed of? (G)

Research Sources

Ⓐ Your campus facilities or physical plant department may be able to tell you where wastewater and storm runoff go, how it is treated, and how much is generated.

Ⓑ The campus development office or environmental health department may have information on methods to estimate sewage generation. Environmental health can also tell you what agencies oversee or regulate wastewater and storm runoff, and whether anyone inspects your campus discharge for contaminants.

Ⓒ The local (probably municipal) wastewater treatment plant can tell you about the treatment processes provided, plant capacity, and any regulations pertaining to sewage discharge.

Ⓓ Your city, county, or state health department can give you information about health standards. Regulations regarding gray water reuse differ from city to city and state to state. In many areas the use of gray water is not legal or requires a special permit or variance.

Ⓔ Your regional water-quality agency (possibly state) or EPA will have information on regulations, inspection, and local water-quality issues.

Ⓕ Local groups concerned with ocean or lake pollution may have information on local treatment plants.

Ⓖ Your agriculture extension (county) or the agricultural department on campus can tell you about water used for crops and livestock.

Ⓗ Your city's environmental or conservation office may have information on water use and wastewater-reduction programs.

Recommendations

Reducing wastewater volume and toxicity are important goals in any effective wastewater-management program. You can help make the following priorities a reality on your campus:

✅ Establishing water conservation programs to reduce wastewater generation (see "Water," page 45).

✅ Promoting policies which encourage the campus to pay the full charge for wastewater treatment. Your school may be partially exempt from such charges if it is a state university. These measures will act as incentives to reduce flows.

✅ Exploring the possibility of constructing on-site wastewater treatment plants for new campus facilities.

✅ Constructing gray water systems which capture water from acceptable sources and divert it for subsurface landscape irrigation. Because all hazardous substances need to be kept out of gray water systems, managers will need to select gray-water sources carefully.

✅ Exploring the possibility of using local reclaimed wastewater on campus for landscaping and other nonpotable water uses.

✔ Educating the campus community to minimize drain disposal of chemicals and the use of toxic substances in automotive shops, in research labs, in the classroom, and in janitorial services.

✔ Washing campus vehicles in a central location with catch basins to capture dirt and grease for proper disposal.

✔ Reducing the amount of impermeable surface on campus (asphalt and concrete) by planting trees and increasing green space.

✔ Using dry "vacuum" machines to clean sidewalks and plazas.

RESOURCES

Organizations, Institutions, Contacts

Environmental Defense Fund, 257 Park Avenue South, New York, NY 10010; 1-800-CALL-EDF.

Heal the Bay, 1640 Fifth Street, Suite 112, Santa Monica, CA 90401; (310) 394-4552.

National Estuaries Program, U.S. Environmental Protection Agency, Coastal Management Branch Office of Wetlands, Oceans, and Watersheds, 499 S. Capital Street, Fairchild Building, Suite 811, Washington, D.C. 20460; (202) 260-6502.

National Toxics Campaign, 37 Temple Place, Boston MA 02111; (617) 482-1477.

Natural Resources Defense Council, 40 W. 20th St, New York, NY 10011; (212) 727-2700.

Office of Water Reclamation, Room 366, City Hall, 200 North Spring Street, Los Angeles, CA 90012; (213) 237-0887.

United Nations Environment Programme, UN Plaza 2, Rm. DC2/0803, New York, NY 10017; (212) 963-8093.

WateReuse Association of California, 915 L Street, Suite 1000, Sacramento, CA 95814-3701; (916) 442-2746.

CHAPTER 6:
PEST CONTROL

Pest control presents hazards both to pest control applicators and to students and staff. On many campuses, chemical pesticides are used more often indoors than outdoors. Pesticides may cause health problems, such as headaches and rashes, especially for those who apply the pesticides and those who live or work in treated buildings. Also, pesticides may contaminate irrigation runoff. The Federal Insecticide, Fungicide and Rodenticide Act (FIFRA) regulates the production, use of, and exposure to pesticides. Individual states may have stricter guidelines. Pesticides may be regulated by the department of health, agriculture, or a statewide environmental agency. Some institutions are turning to integrated pest management—reducing chemical use and using less-toxic chemicals—as well as nonchemical means of pest control to reduce the use of toxic chemicals on campus.

In studying the pesticide procedures on your campus, make sure you understand the following terms first: integrated pest management, pesticide, herbicide, fungicide, rodenticide.

Assessment Questions

◆ What kinds of pests are of concern on your campus? Why? Are some areas of campus more prone to pests than others (e.g., agricultural areas, food services)? (See "Research Sources:" A,B)

◆ Does your campus have a pest-control staff? Are services contracted to a private firm? (A)

◆ What pesticides are used on campus and in what quantities? Who manufactures these chemicals? Are more pesticides used indoors or outdoors? Have amounts increased or decreased over the last several years? Why? (A,B)

◆ Are pesticides used that are known or suspected to be carcinogens? In what quantities and locations are these substances used? Are alternatives being considered? (A)

◆ What public agency oversees the use of pesticides on campus? Does it inspect your campus? Has your campus been cited for improper storage or use? Can you compare this information with the records of similar institutions? (A,D,F)

CAMPUS PROFILE:
Georgetown University, Washington, D.C.

Pests or Pesticides: Which are Worse?

The landscape superintendent at Georgetown University is committed to reducing the number of harmful substances in the campus environment. After taking classes on integrated pest management at the University of Maryland, the superintendent applies the principles he has learned directly to the campus. When an insect population is found on the grounds, instead of "going to a manual and dialing up the hottest pesticide," members of the grounds crew look at the life cycle of the species to locate its weak points. They use nontoxic soaps and oils to interrupt the insects' lifecycles—by reducing mating, for example. They allow beneficial insects to live; and rather than spraying pesticides, they dump out, move, or recycle cans that attract wasps and other insects, whenever possible. Although the campus still hires an herbicide company to kill weeds, they are also trying out another firm that treats pest problems individually instead of blanket spraying, and that uses products that are more rapidly biodegradable.

◆ How are pesticides applied? Are warning signs posted giving information about time and place of pesticide application? Do these warnings meet or exceed state and federal regulations? How are excess pesticides disposed of? (A,C,G)

◆ Are pesticide applicators warned of potential hazards of such chemicals and are they given proper training and protective equipment? (C,G)

◆ Are nonchemical pest-control methods used on campus? What methods are used and in what locations? With what results? Are nonchemical methods being considered for future use? (A,F)

◆ Does your school have agricultural areas? Does it operate a ranch, farm, or campus garden? If so, what pest control methods and products are used? (E)

◆ Is there a possibility of groundwater contamination or contaminated runoff from agricultural or landscaped areas? (D,E,F)

◆ Is your campus part of a larger state, county, or municipal pest-control project (such as mosquito abatement)? (D,F)

◆ What is the cost of campus pest control? Who pays these costs? (A,B)

Research Sources

🅐 Your campus maintenance office is probably responsible for pest control. This is your best source for finding out what pest-control methods are used on campus, the quantities of pesticides used, application methods and schedules, warning procedures, and costs.

🅑 The campus purchasing office may also have information on the amount of pesticides ordered and the cost of pest control.

🅒 Your campus environmental health office should have information on the disposal of chemical pesticides and personal protection for applicators.

🅓 The county, state, or regional agency that oversees pesticide use in your area should be contacted for local regulations and inspection records. They should have a list of all chemicals used on campus landscapes. Specific precautions on the use of these substances will be contained in these files. Who monitors their use? What are the monitoring procedures?

🅔 Your campus agricultural department and the cooperative extension (also called "agricultural extension") associated with most county governments can provide you with information about pest-control practices and technologies used in your area.

🅕 Local environmental groups and state and local agencies may have information on groundwater contamination and nonchemical pest-control methods.

🅖 Your local employee union can provide information regarding pest-control application safety and training.

Recommendations

The policies and practices regarding pesticide on your campus may involve larger governmental entities than your campus administration. Try to involve your community representatives when you discuss the following:

✔ Reducing the use of chemical pesticides and exploring the use of organic, biological, and other new technologies instead.

✔ Eliminating the use of all pesticides listed by the EPA and your cooperative extension as carcinogenic.

✔ Educating the campus community about keeping areas clean to avoid attracting pests.

✔ Keeping landscaped areas healthy, by organic means, to increase their natural resistance to disease and pests.

✔ Following recommended guidelines for proper protection of pesticide applicators and proper warnings to those who use them.

❖ Encouraging Integrated Pest Management (IPM) strategies, which rely on biological and cultural controls and minimize synthetic chemical use (see "Research Sources," page 30).

RESOURCES

Books, Publications, Articles

Carson, Rachel, *Silent Spring*. Boston: Houghton Mifflin, 1987.

Common Sense Pest Control Quarterly. Newsletter published by the Bio-Integral Resource Center, P.O. Box 7414, Berkeley, CA 94707.

The IPM Practitioner. Monthly newsletter of the Bio-Integral Resource Center (see above).

Journal of Pesticide Reform. Northwest Coalition for Alternatives to Pesticides, P.O. Box 1393, Eugene, OR 97440.

Shepherd's Purse. Pest Publications. Summertown, Tenn.: 1987.

Yepsen, Roger B, Jr. *The Encyclopedia of Natural Insect and Disease Control*. Emmaus, Penn.: Rodale Press, 1984.

Institutions & Organizations

Agroecology Program, University of California, Santa Cruz, CA 95064; (408) 429-4140.

Bio-Integral Resource Center, P.O. Box 7414, Berkeley, CA 94707; (510) 524-2567.

National Coalition Against the Misuse of Pesticides, 710 East St., S.E., Suite 200, Washington, D.C. 20003; (202) 543-5450. Contact for information on toxic-free pest control (indoors and outdoors).

National Pesticides Telecommunications Network, 1-800-858-PEST.

Northwest Coalition for Alternatives to Pesticides, P.O. Box 1363, Eugene, OR 97440; (503) 344-5044.

University of California Statewide Integrated Pest Management Project, Education and Publications Office, Davis, CA 95616; (916) 752-8876.

U.S. Environmental Protection Agency (EPA), Office of Pesticide Programs, Crystal Mall Building 2, Room 1115, 1921 Jefferson Davis Highway, Arlington, VA 20460; (202) 557-5725.

U.S. Public Interest Research Group (USPIRG), 215 Pennsylvania Ave., S.E., Washington, D.C. 20003; (202) 546-9707.

CHAPTER 7:
AIR QUALITY

Air pollution is perhaps the most pervasive environmental problem affecting people's health. More than 100 million Americans live in areas that do not meet federal health standards for such pollutants as carbon monoxide and smog. Statistics are scarce on how many people are exposed to hazardous levels of toxic air pollutants, a group of substances that have only recently begun to be studied. On a global scale, air pollution contributes to the Greenhouse effect, which, if continued unabated, is predicted to lead to a drastic warming of the planet. In addition, various substances—primarily chlorofluorocarbons—are contributing to the depletion of the earth's protective ozone layer.

Campuses can contribute to air-quality problems—and their solutions—in a number of ways. Three types of air pollutants need to be considered:

❶ Criteria pollutants: carbon monoxide, nitrogen oxides, sulfur oxides, reactive organic gases, lead, and airborne particulates (primarily generated by carbon combustion).

❷ Toxic air pollutants (primarily generated by industrial manufacturing).

❸ Ozone-depleting substances (primarily generated by refrigeration and air conditioning).

Before you begin your audit, you will need to review your knowledge of the Greenhouse effect and ozone-layer depletion, and be prepared to discuss the following: freon recovery, chlorofluorocarbons, ozone-depleting compounds, halons, criteria pollutants, toxic air pollutants.

Assessment Questions

◆ What is the ambient air quality for criteria pollutants in the area? Are federal or state standards exceeded? Which ones? How often? (See "Research Sources:" C,E,F)

◆ What sources on campus emit toxic air pollutants? Probable sources include steam plants, paint booths, incinerators, laboratory fume hoods, fuel dispensing, and fleet and private vehicles (see "Transportation," page 64). (A,B,C,D,F)

◆ What kinds of ozone-depleting compounds are used on campus and what are the sources? These substances may be released during servicing of refrigeration and air conditioning equipment (in both autos and buildings), during sterilization of medical equipment, and from museum documents sprayed with protective chemicals. How much of each pollutant is emitted? (A,B,E)

◆ What is being done on your campus to control and reduce air emissions? Identify emission-control devices. What is required? What is in place? What is available? Are alternative, cleaner fuels (methanol or natural gas, for example) used where possible? What is the potential for using such fuels on campus? Have renewable energy sources (solar, wind, etc.) been considered as additional alternatives? (A,C)

EDUCATION ENVIRONMENT ENLIGHTENMENT

CAMPUS PROFILE:
University of Kansas, Lawrence

Breathing Easier

In the fall of 1990, the University of Kansas established an Environmental Ombudsman's Office. This office is now responsible for researching campus environmental impacts, and implementing programs that contribute to improving environmental quality as well as reducing university costs. One of its most successful programs is the school's Freon Recovery Project, which, in its first six months of existence, recaptured 3,400 pounds of freon, about one-third of the campus's total purchase. This resulted in a savings of $15,000, well over the amount needed to start the program.

The project is part of an overall campus effort to reduce the use of ozone-depleting compounds, which, in addition to freon, include chlorofluorocarbons used in refrigeration and air conditioning, halons used in fire extinguishers, carbon tetrachloride used in laboratories, and chemical sprays used to preserve maps and museum documents. A committee comprised of representatives from the campus air-conditioning shop, the garage, facilities planning, the Housing Department and the Environmental Health and Safety Office is compiling a list of ozone-depleting compounds used on campus and is investigating alternatives and minimization strategies.

Additionally, the Environmental Ombudsman's Office is in the process of reviewing other campus environmental impacts and developing comprehensive conservation and waste-reduction programs. These include a university-wide pesticide policy, expanded campus recycling and procurement of recycled materials, improved energy efficiency with the use of compact fluorescents and light-management plans, and water-conservation strategies.

◆ Are any programs in place to recapture ozone-depleting compounds or find safer alternatives? (A,B)

Research Sources

Ⓐ Campus physical plant personnel may provide information on campus sources and emissions. Your school may report emissions from some stationary sources (steam plants, for example) that may require operating permits from local air-pollution control districts. Rather than being measured, emissions may be estimated using standard emission factors (tons of pollutant per 1000 gallons of fuel burned). You may be able to estimate freon (CFC) releases from the amounts purchased. Many sources, particularly toxic pollutants, cannot be quantified accurately.

Ⓑ The users of ozone-depleting compounds should be queried. These include the campus garage, housing department, facilities, and other campus maintenance departments.

Ⓒ Local air-pollution control agencies should be able to provide information on permits, emissions reporting, emissions factors, and local air quality.

Ⓓ Environmental impact reports for campus construction projects may have information on emissions.

Ⓔ The U.S. Environmental Protection Agency (EPA) regional office may be able to provide information on ambient air-quality and emissions factors. Check the regional office library, if available.

Ⓕ Your state government's regional EPA office can provide state and federal "right-to-know" regulations, which may require reporting of toxic pollutant releases.

Recommendations

Campus air quality may be affected by a constellation of sources and operations. Involve as many responsible parties as possible in your discussions of the following:

✔ Improving the monitoring of emissions of stationary sources of pollution, and helping design control strategies.

✔ Improving the control technology of stationary pollution sources.

✔ Using cleaner fuels, such as natural gas, in steam plants and watercoolers and chillers.

✔ Encouraging the use of alternative-fuel and electric service vehicles on campus.

✔ Exploring the use of renewable energy, such as solar and wind energy, to supply campus power needs, thus reducing the air emissions caused by fuel-burning energy plants.

✔ Reducing vehicle use in general (see "Transportation," page 64).

✔ Reducing the use of toxic substances throughout the campus.

✔ Installing CFC-recycling and recovery equipment for use in vehicles, air conditioning, and refrigerator maintenance.

✔ Encouraging the use of alternatives for ozone-depleting compounds used on campus.

RESOURCES

Books, Publications, Articles

MacKenzie, James, *Breathing Easier: Taking Action on Climate Change, Air Pollution, and Energy Insecurity*. New York: World Resources Institute, 1988. 1750 New York Avenue, N.W., Washington, D.C. 20006; (202) 638-6300.

U.S. EPA, Office of Air and Radiation. *How Industry is Reducing Dependence on Ozone-Depleting Chemicals,* June 1988.

U.S. EPA, Office of Air and Radiation, Office of Air Quality Planning and Standards. *Toxic Air Pollutant Emissions Factors: A Compilation for Selected Air Toxic Compounds and Sources,* by Anne A. Pope and Patricia A. Cruse. Research Triangle Park, N.C.: EPA-450/2-88-006, October 1988.

Institutions & Organizations

Coalition for Clean Air, 122 Lincoln Blvd., Suite 201, Venice, CA 90291; (310) 450-3190.

Cool It! National Wildlife Federation, 1400 16th Street, N.W., Washington, D.C. 20036; (202) 797-5435.

Environmental Ombudsman's Office, Susan Ask, Associate Environmental Ombudsman, University of Kansas, Hayworth Hall, Biology, Lawrence, KS 66045; (913) 864-3208.

Greenhouse Crisis Foundation, 1130 17th Street, Ste. 630, Washington, D.C. 20036; (202) 466-2823.

Natural Resources Defense Council, 1350 New York Avenue, N.W., Washington, D.C. 20005; (202) 783-7800.

CHAPTER 8:
THE WORKPLACE ENVIRONMENT

Environmental health hazards in the workplace are drawing increasing public attention. Numerous substances found on campus may expose students, faculty, and staff to a variety of health risks. These substances include cleaning and pest-control products, art, photography and theater materials, chemicals used in the classroom and research, pollutants in drinking water, asbestos, radon, toxic building materials, biological contaminants, and tobacco smoke. Without adequate precautions, exposure to these substances and materials may cause health effects as minor as headaches or as major as respiratory problems, nervous system disorders, asthma, allergies, immune disorders, and even cancer. Under the federal Occupational Safety and Health Act, employers are required to provide information to workers regarding exposure to hazardous substances on the job. These regulations, however, do not apply to students. Also, many commonly used substances on campus, such as cleaning products, art and photographic supplies, and building materials, may not pose health risks alone, but combined with other substances, or used in the presence of poor ventilation, they may contribute to poor indoor air quality, which can impact health. Substituting safer products, implementing education and safety-training programs, and improving building maintenance can all help reduce environmental health risks found on campus.

You may wish to poll a variety of areas on campus and evaluate the safety standards of each. Background study should include the range of substances in each location that could cause health or safety problems.

Assessment Questions

◆ What kinds of occupational environmental hazards may exist on your campus? Does your school have a history of worker/student health and safety complaints? (See "Research Sources:" A,B,C,F)

◆ What are the environmental health and safety policies on your campus? Does the state system to which your campus belongs have such policies? Do they apply to students or only to employees? (A,C)

CAMPUS PROFILE:
University of the Arts
Philadelphia, Pennsylvania

Safe Art

At art and design schools, where toxic materials are a daily part of campus life, health and safety issues can be critical. A survey conducted by students in 1991 at the University of the Arts as part of a campus environmental audit revealed startling information about the lack of safety procedures governing the use and disposal of hazardous art supplies. Less than half of the students answering the questionnaire said that their department had a safety procedure. Only 32 percent said that these policies were enforced and less than half of the respondents said that they had been thoroughly informed of the hazards associated with the art materials they use. The survey showed that inspections for safe working conditions were sporadic and ventilation was frequently inadequate. The survey also indicated that many students would consider using safer alternatives, would properly dispose of hazardous art material if it were more convenient, and would support a mandatory course for health and safety.

Based on the survey results, students developed a comprehensive proposal for a health and safety program for the College of Art and Design, which has been presented to the administration by students, the director of Public Safety, and the school nurse. The proposal outlines numerous steps to ensure adequate health and safety precautions. These include training seminars for faculty and shop supervisors, student workshops, tailoring health and safety programs to meet departmental needs, monthly inspections, freshman orientation programs, increasing the availability of first-aid supplies and protective equipment, and training seniors how to set up an environmentally sound studio after graduation. To measure progress and revise the program, the proposal also requests that the administration conduct an audit of the school every five years.

◆ What office handles occupational environmental health and safety? Is there an emergency response team on campus? Is there an evacuation plan in place? Is there an accident-prevention program? (A)

◆ Is your campus located in an area subject to environmental safety hazards such as earthquakes, hurricanes, tornadoes, or other natural disasters? Are appropriate safety precautions and emergency plans in place? Are they periodically tested? (A,D)

◆ What job classifications on campus require the handling of special hazards (e.g., asbestos, medical waste, PCBs, or pesticides)? Do employees in these jobs receive special safety training? (A,B,D)

◆ Is occupational safety training required? If so, how often? Is this training required for specific positions only or for all employees? Is occupational health and safety training required and/or available for students and faculty in certain high-risk departments (e.g., chemistry, medicine, art)? (A,B,C)

◆ What is the level of compliance with environmental health and safety policies on your campus? How is the policy enforced (if at all)? How available is occupational health and safety information on campus? Do individual departments have their own safety committees and/or policies? How are such policies enforced? (A,E)

◆ Are facilities inspected for safe working conditions? By what agency? Are citations issued for unsafe conditions? Has your campus received any citations? (A,E)

◆ Are any courses or seminars conducted on art safety for students? Are faculty trained in art safety? (A,C)

◆ What is the quality of ventilation in the areas where art materials and photographic chemicals are used? Are inspections of these areas conducted on a regular basis? How do the monitoring procedures differ for art departments versus science departments? (A,C,G)

◆ Is there a nonsmoking policy on your campus? (A)

◆ Have workers or students complained of health problems that may be due to poor indoor air quality? What are the nature of the complaints? Symptoms such as headaches, drowsiness, lightheadedness, and nausea, for example, may be caused by sources such as poor ventilation, new carpets and furniture, paints, varnishes, adhesives, or tobacco smoke. Have campus officials investigated possible indoor air-pollution problems? (A,B,C,E)

◆ Have radon levels been monitored on campus? What is the status of asbestos abatement in campus facilities? What is the quality of drinking water on campus? Is water quality a concern in your community? Has it been tested for contaminants, such as lead? (A,E,G)

◆ Has your school developed policies and procedures regarding the use of video display terminals (VDTs)? (A,B,C)

◆ Is your campus located near a power-generating station? Is there any concern on campus regarding electromagnetic fields (EMFs)? (A,E,F)

Research Sources

◆ Your campus office of environmental health should have information on campus safety policies, programs, and training. They will also know what agency inspects the campus for safety violations. Check with individual departments for safety information as well.

B Unions representing service staff can tell you which safety and health issues employees are concerned about, and what training programs are available for employees.

C Students, faculty, and chairpersons of various departments are good sources of information.

D The personnel office can provide job classifications and training information.

E The agency that inspects working conditions on campus (perhaps your state/local OSHA office) can provide inspection reports. You may also want to get similar information from other local campuses for comparison.

F Your state's department of health services should be contacted for additional information on environmental health hazards in the workplace.

G Your local yellow pages under "environmental testing" has a list of certified labs and consulting services that can test samples of workplace toxins.

Recommendations

Establishing goals that are appropriate to each department will be key to improving the workplace safety of your campus. You can design targeted surveys for specific campus populations, such as students and staff who regularly use toxic substances in their work. Try to find one or two people in each department who can maintain the policy as well as help develop it in their areas. Use the following ideas as starting points:

✔ Surveying student and employee safety concerns to identify problem areas that require special attention.

✔ Appointing safety officers for each building and/or department to coordinate safety information and training.

✔ Expanding training programs and exploring ways to create or strengthen health and safety policies to govern the handling of materials not normally considered hazardous, such as art and architecture supplies, cleaning products, etc.

✔ Identify individual staff members, students, or create a committee to serve as a liaison to unions on your campus and help them improve their workplace environment.

✔ Establishing no-smoking policies in all buildings. These policies should be building-wide. Even if one department or floor prohibits smoking, tobacco smoke can be circulated to other offices through ventilation systems.

✔ Duplicating effective safety programs already in place in one department for use in other departments.

✔ Encouraging the creation of art safety classes which link environmental issues with art and design and train the users of art, photography, and crafts materials.

✔ Properly maintaining and regularly monitoring heating, ventilation, and air-conditioning (HVAC) systems campus-wide.

✔ Testing for and mitigating pollutants such as radon, asbestos, and drinking water contaminants.

✔ Reducing the use of substances which may contribute to poor indoor air quality, such as pesticides and cleaning products.

✔ Using nontoxic building materials wherever possible in new building construction and renovation (see "Campus Design and Growth," page 68).

✔ Developing policies and procedures for VDT use that identify ergonomic design features, specify frequency of breaks, etc.

RESOURCES

Books, Publications, Articles

Dadd, Debra Lynn. *Nontoxic, Natural, & Earthwise*. Los Angeles: Jeremy P. Tarcher, Inc., 1990.

Evaluating Office Environmental Problems, Annals of the American Conference on Governmental Industrial Hygienists, vol. 10. Akron: American Industrial Hygiene Association, P.O. Box 8390, Akron, OH 44320. Oct. 1984.

Gay, Kathlyn. *Silent Killers*. New York: Franklin Watts Inc., 1988.

The SEAC Organizing Guide, SEAC, Chapel Hill, NC, 1990.

Understanding Toxic Substances: An Introduction to Chemical Hazards in the Workplace. 1986. Labor Occupational Health Program (LOHP), University of California, 2531 Channing Way, Berkeley, CA 94720; (510) 642-5507.

U.S. EPA, Office of Air and Radiation. *The Inside Story: A Guide to Indoor Air Quality*. Washington, D.C. 20460. Sept. 1988, EPA/400/1-88/004.

Institutions & Organizations

Action on Smoking and Health (ASH), 2013 H Street N.W., Washington, D.C. 20006; (202) 659-4310.

Americans for Nonsmokers' Rights, 2530 San Pablo Avenue, Suite J, Berkeley, CA 94702; (510) 841-3032.

Clean Water Action, 317 Pennsylvania Ave, S.E., Washington, D.C. 20003; (202) 547-1196.

Group Against Smokers' Pollution (GASP), 25 Deaconess, Boston, MA 02115; (617) 266-2088.

Labor Occupational Safety and Health (LOSH) Program, University of California, Los Angeles, 1001 Gayley Avenue, 2nd floor, Los Angeles, CA 90024; (213) 825-7012.

National Center for Environmental Health Strategies, 1100 Rural Avenue, Voorhees, NJ 08043; (609) 429-5358.

National Institute for Occupational Safety and Health (NIOSH), Hazard Evaluations and Technical Assistance Branch (R-9), Division of Surveillance, Hazard Evaluations and

Field Studies, U.S. Department of Health and Human Services, 4676 Columbus, OH 45226.

9to5, National Association of Working Women, 614 Superior Ave., N.W., Cleveland, OH 44113; (216) 566-9308.

Occupational Safety and Health Administration, U.S. Department of Labor, 200 Constitution Ave, N.W., Washington, D.C. 20210; (202) 523-6091.

Safe Buildings Alliance, Metropolitan Square, 655, 15th St., N.W., Suite 12, Washington, D.C. 20005; (202) 879-5120. For information about indoor air problems.

SEAC, P.O. Box 1168, Chapel Hill, NC 27514; (919) 967-4600. SEAC has many contacts with many labor unions on college and in communities across the country.

Toxic Substance Control Act Assistance Information Service; (202) 554-1404. For information on EPA's asbestos programs.

ART HAZARDS

Books, Articles, Publications

Art and the Craft of Avoidance, 1991. U.S. Public Interest Research Group, 215 Pennsylvania Avenue, S.E., Washington, D.C. 20003; (202) 546-9707.

Art Hazard News. Newsletter of the Center for Safety in the Arts, 5 Beekman Street, Suite 1030, New York, NY 10038; (212) 227-6220.

McCann, Michael. *Artist Beware: The Hazards and Precautions in Working with Art and Craft Materials.* New York: Watson-Guptill Publications, 1979.

McCann, Michael. *Health Hazards Manual for Artists.* New York: Lyons & Burford, 1985.

Qually, Charles. *Safety in the Artroom.* Davis Publications, 1986.

Rossol, Monona. *The Artist's Complete Health and Safety Guide,* 1990. Allworth Press, 10 East 23rd St., New York, NY 10010; (212) 777-8395.

The Safer Arts. Health and Welfare Canada, Communications Branch, Publications Distribution Center, 19th floor, Jeannae Mance Building, Tunney's Pasture, Ottowa, CANADA KIAOK9; (613) 952-9190.

Institutions & Organizations

The Arts and Craft Materials Institute, 715 Boylston Street, Boston, MA 02116; (617) 266-6800.

Center for Safety in the Arts, Center for Occupational Hazards, 5 Beekman Street, Suite 1030, New York, NY 10038; (212) 227-6220. A national clearinghouse for information on hazards in the arts. Write or call for a complete list of publications.

College Art Association, 275 7th Avenue, 5th Floor, New York, NY 10001; (212) 691-1051.

National Association of Schools of Art and Design, 11250 Roger Bacon Drive #21, Reston, VA 22090; (703) 437-0700.

University of the Arts, Public Safety Office, 320 S. Broad, Furness Hall, Philadelphia, PA 19102; (215) 875-4804.

U.S. Public Interest Research Group, 215 Pennsylvania Avenue, S.E., Washington, D.C. 20003; (202)546-9707.

SECTION II:
RESOURCES AND INFRASTRUCTURE

YOUR CAMPUS IS A MICROCOSM OF THE *larger community that surrounds it. Every campus ecosystem has an infrastructure that supports the housing, transportation, and feeding of its resident population. How your campus manages its energy and water resources, what products its food service buys, and what kinds of paper and other goods its offices and labs use can have a profound effect on the local and regional environment. Those campuses that develop effective "green consuming" policies can serve as models for the community that lies beyond the campus gates. As residents of the campus community, students have an opportunity and a responsibility to help manage the resource flow on their campuses.*

SECTION II: RESOURCES AND INFRASTRUCTURE

Water use by individuals and institutions is not generally regulated, even though many parts of the country are experiencing droughts or water shortages. Regardless of your region's climate, it is important to conserve water, as groundwater supplies are increasingly depleted and polluted. And by cutting back on the volume of wastewater and runoff generated by your campus you will also be helping to cut back on the amount of pollutants entering your local waterways and regional bodies of water. Water policy may be set at the state or local level; there are no federal guidelines for water consumption. Your state or local authority may have voluntary or mandatory water conservation guidelines. Both indoor and outdoor consumption are significant, though outdoor use is more heavily influenced by the local climate. Water conservation measures can significantly reduce the amount of water your campus uses.

Before beginning your survey, find out about low-flow water devices, xeriscaping, and drip irrigation.

Assessment Questions

◆ Where does your campus water supply originate? What is the status of that supply? Is the water table in danger of being depleted? (See "Research Sources:" E,F)

◆ How many gallons of water does your campus use annually? (Water use is usually available as the number of billing units, which must be converted to gallons based on the supplier's formula. Measurements are sometimes given in acre-feet. There are 326,000 gallons in 1 acre-foot.) Convert this to per-capita and per-square-foot consumption. Has this figure increased, decreased, or remained constant over the past five years? Why? (A)

◆ What percentage of water is used indoors versus outdoors? (A)

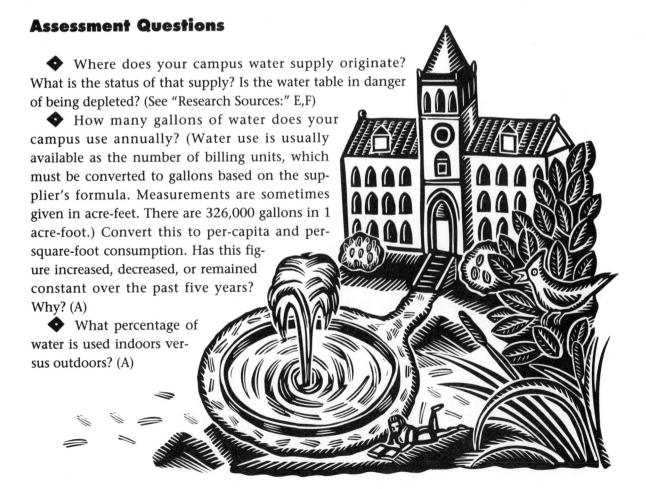

CAMPUS PROFILE:
California State University, Northridge

Campus Drought Busters

Cal State Northridge (CSUN) has been actively involved in water conservation since the California drought of 1977. During 1990, CSUN achieved a reduction of 14 percent from the base year of 1986. Since then, the university has taken steps to reduce consumption an additional 10 percent in anticipation of worsening drought conditions. Water-conserving measures already in place include retrofitting all showers, flush valves, and faucets on campus with water-saving devices, reducing and changing irrigation schedules, posting water-conservation stickers in all restrooms and kitchenettes, eliminating washing of university vehicles, and distributing educational materials. CSUN continues to explore new methods of water conservation, such as replacing obsolete and malfunctioning irrigation systems and using reclaimed water for landscaping purposes.

◆ Is water use metered separately for individual buildings? If so, which buildings use most? Is the irrigation system metered? Is meter information used to reduce water use? (A)

◆ Does your campus have a system to monitor for leaks or water efficiency? (A)

◆ What type of irrigation system does your campus use? Automatic sprinklers, timers, weather-related management and drip irrigation will help reduce consumption. (A)

◆ Has a campus water-conservation program been implemented? When and why was the program established (increased cost of water, community pressure)? What measures are included? How has it altered water-use patterns? (A,D)

◆ Do campus buildings have low-flow water devices for toilets and sensors for faucets? Have older buildings been retrofitted? What standards exist for new buildings? (A,C,D)

◆ How much water does campus landscaping require? Is it adapted to the local climate? Is it drought-tolerant? (A,C)

◆ Is reclaimed water being used in any facilities or for landscaping purposes? (A,C)

◆ If you live in a drought-prone area, have students, faculty, and staff been informed about methods of water conservation? Are public notices posted in bathrooms and other locations to encourage saving water? (A,D)

◆ Who pays for water use? Is each department billed separately or is one water bill generated for the entire campus? (A,B)

◆ What is the school's annual water expense? Has the amount increased or decreased over the past five years? Are changes the result of usage factors, changes in the cost of water, or both? How have population increases and expansion of facilities affected water use? (B)

◆ Does your campus have agricultural areas? How much water do these areas use? (A,D)

◆ How does water use at your campus compare to other institutions for which statistics are available? (E)

Research Sources

Ⓐ Campus facilities or maintenance departments may be able to provide information regarding water use, monitoring, and inspection of facilities. They may also have information on water meters, irrigation systems, and indoor plumbing fixtures. Ask them about campus water-conservation programs.

Ⓑ The campus facilities or accounting office may be able to tell you about billing procedures, the amount spent on water, and who the local water supplier is. Ask for quantities for the last several years in order to chart any change in water use.

Ⓒ The planning or development office may have plans for future campus expansion. These should include a discussion of water use.

Ⓓ Individual academic departments—including agriculture, horticulture, and engineering—may have information on water-saving fixtures, landscaping methods, and other water-conservation programs available to your campus.

Ⓔ Consult local and regional water suppliers and policy groups for an understanding of water-supply and water-quality issues in your area. The local supplier may be able to give you consumption figures for other mixed-use institutions in your region to allow per-capita comparisons.

Ⓕ Local environmental groups as well as state and local agencies may have information on groundwater and surface water-supply depletion.

Recommendations

Water conservation can be an individual as well as a community effort. Students, faculty, and staff should be encouraged to implement water-conserving methods at home as well as on campus:

✔ Installing water-conserving plumbing fixtures (shower heads, toilets, and faucets) in all new buildings. Retrofitting older buildings.

✔ Installing a water-conserving irrigation system, such as drip irrigation.

❤ Landscaping with drought-tolerant plants ("xeriscaping") and planting native species wherever possible.

❤ Modifying irrigation schedules to reduce water use (by watering in the early morning and late evening, when evaporation is lowest, for example). Checking irrigation systems regularly for malfunctions.

❤ Exploring the feasibility of billing individual departments for water use in order to encourage conservation and discourage waste.

❤ Implementing comprehensive leak-detection and maintenance programs.

❤ Educating the campus community about water-conservation measures. Posting "Save Water" stickers in all restrooms and kitchenettes.

❤ Eliminating or severely curtailing the washing of fleet vehicles.

❤ Exploring the feasibility of using reclaimed water for irrigation and other nondrinking water uses (see "Wastewater and Storm Runoff," page 24).

RESOURCES

Books, Publications, Articles

Martin, William, et al., *Saving Water in a Desert City,* Resources for the Future, Washington, D.C.: Johns Hopkins University Press, 1984.

Slater, William, and Peter Orzechowski. *Drought Busters: 30 Easy Ways to Save Water...and Money.* Venice, Calif.: Living Planet Press, 1991.

Institutions & Organizations

Brown is Green, Box 1941, Brown University, Providence, RI 02912; (401) 863-7837.

California State University, Northridge, Water Conservation Program, Physical Plant, Northridge, CA 91330; (818) 885-4630.

Environmental Defense Fund (EDF), 257 Park Avenue South, New York, NY 10010; 1-800-CALL EDF.

National Water Center, P.O. Box 264, Eureka Springs, AR 72632; (501) 253-9755.

Natural Resources Defense Council (NRDC), 40 W. 20th St, New York, NY 10011; (212) 727-2700.

Rocky Mountain Institute, 1739 Snowmass Creek Road, Snowmass, CO 81654; (303) 231-7303.

Safe Drinking Water Hotline, (202) 382-5533. For information on regulations under the Safe Drinking Water Act, radon in drinking water, and a list of state drinking water offices.

South Florida Water Management District, P.O. Box 24680, W. Palm Beach, FL 33416; (407) 686-8800. Contact for a free color brochure on xeriscaping.

America's dependency on petroleum contributes to critical environmental problems, such as global warming, acid rain, and smog. Reliance on foreign oil threatens national security, and extracting fossil fuels domestically continues to threaten fragile ecosystems and create the potential for devastating oil spills. Energy efficiency is the least polluting, lowest-cost energy resource available. Cheaper and safer than extracting fossil fuels and constructing dangerous and inefficient nuclear power plants, using energy wisely not only reduces environmental impacts but also saves money. Campus energy-efficiency programs and advanced technology programs have become models for institutional efficiency and have cut millions from utility bills. Using cleaner fuels, such as natural gas, can also reduce environmental impacts significantly.

Before beginning your study, make sure you have an understanding of the different types of energy sources used in your region, as well as the differences between renewable and nonrenewable sources. Other terms include: compact fluorescents, efficient ballasts, room-occupancy sensors, daylighting, photovoltaics, cogeneration, passive solar, building envelopes, and fuel cells.

Assessment Questions

◆ What are the principal sources of energy for your campus and the power plants serving your region: oil, coal, natural gas, methane, propane, hydroelectric, solar, wind, nuclear, or a combination of these? (See "Research Sources:" A,B)

◆ How much energy is used on campus? Units include KWHs of electricity, BTUs or therms of natural gas, and gallons or BTUs of fuel oil. How much energy is used per square foot of maintained building space? This standardized formula will allow you to differentiate between changes in consumption due to growth over time, changes due to increased activity or consumption in existing space, and changes in fuel sources for heating and cooling. You can also convert data to energy use per capita based on campus population statistics. (A)

◆ What are the trends in energy use on your campus? Get existing annual data that has been gathered over the past five years, and monthly data for at least one year. How has usage changed and by how much? Data on individual buildings can be useful, but often it is not available. Weather data (days when heating or cooling is used) may explain some changes in consumption. Data from other campuses in the area may be used for comparison. (A)

◆ What energy-conservation measures have been implemented at your school? What measures are being considered or planned? Distinguish between technical solutions and those requiring changes in human behavior. Technical changes are more certain to produce results, but motivational efforts should not be overlooked. (A)

◆ Are new buildings designed to maximize energy efficiency? Have energy audits been conducted? Has a computerized energy management control system been installed? (A)

◆ What sources of alternative energy are used or could be used? Solar energy might be used to warm buildings or for domestic hot water in residence halls. Explore the use of alternative fuels, photovoltaic energy, cogeneration,wind, and fuel cells, for example. (B,C,D)

◆ What was the school's total energy bill for the past year? (A)

◆ How are energy costs and conservation programs funded? Cost-analysis of each energy source (per unit and per annual totals) may help explain consumption patterns. Are public bond monies or grants available to fund capital costs of energy-conservation measures? Is third-party financing possible? Does your local utility offer an energy rebate program? What pay-back criteria are used for conservation projects? Where do budget dollars saved through conservation measures go? (A,B)

◆ Have you noticed certain buildings that are overheated in the winter or overcooled during warm months? How are thermostats controlled and at what temperatures are they typically set? How frequently are furnaces serviced? (A,E)

◆ How does your school compare with other institutions in respect to energy use per square foot of building space? (A,D)

Research Sources

Ⓐ Campus physical-plant personnel, the energy manager, and the campus planning or development office may be contacted for information about campus energy consumption and conservation.

Ⓑ Local public utilities generally welcome requests for information about conservation programs and incentives, such as rebates. They can also tell you from what sources the electricity they supply is generated.

Ⓒ Experts, such as faculty members, energy consultants, utility representatives, or local and national environmental groups, should be able to provide advice on possible conservation measures and alternative energy and energy-efficient measures.

Ⓓ Other schools, such as Harvard, SUNY Buffalo, and Brown University, have excellent energy programs.

Ⓔ The campus population could be surveyed for information on indoor temperature settings.

Recommendations

A number of measures, both large and small, may be included in your plans to improve energy efficiency on campus. Here's a sample of ideas to discuss:

✔ Improving lighting efficiency (by using compact fluorescent bulbs, reflectors, efficient ballasts, and room-occupancy sensors, for example). Although compact fluorescent lighting is more expensive than incandescent, the cost will pay off through the savings in energy.

EDUCATION ENVIRONMENT ENLIGHTENMENT

CAMPUS PROFILE:
State University of New York at Buffalo

Power-hungry Schools

By analyzing the State University of New York (SUNY), Buffalo's energy-use patterns in the early 1980s, energy officer Walter Simpson and biophysics professor Fred Snell determined that the university produces 313,900 tons of CO_2 emissions each year, or 10.5 tons of this greenhouse gas for every student, faculty, and staff member on campus. According to their research, electricity, which is generated at a coal-fired power plant, accounts for 71 percent of the school's annual carbon dioxide emissions. Emissions generated from commuters were also significant, while maintenance and public-safety operations contributed only a small fraction of the total. Concerned about the consequences of global warming, the researchers made several recommendations in their report to reduce the university's energy consumption. These recommendations included an accelerated campus-wide energy-efficiency program, with particular emphasis on new building design and construction; conversion to natural gas wherever possible; increased faculty research in the areas of energy efficiency and renewable energy sources; cutting less grass on campus lawns and planting more trees; and increasing awareness about global warming both on campus and in the surrounding communities.

Since the completion of their study, SUNY Buffalo launched a comprehensive energy-efficiency program in 1982, which has reduced the university's utility bill by $3 million a year. Under the "Conservation UB" program, substantial accomplishments have been achieved. In the past decade, more than 100 lighting, heating, and cooling projects have been implemented, and other conservation projects—such as the renovation of fume-hood exhaust systems—are expected to dramatically increase savings and reduce CO_2 emissions in the years ahead.

✅ Incorporating passive solar building design and energy efficiency into future building plans, such as the use of "daylighting," a means of maximizing the use of natural light in the design of a building. In addition, planners may be encouraged to incorporate safe and renewable energy sources such as photovoltaics, cogeneration, wind, and fuel cells into campus energy plans.

✅ Increasing the efficiency of heating and cooling systems. Look at heating- and cooling-season temperature policies (a change by only a few degrees can create tremendous savings). Report overheating and overcooling of buildings to the physical plant manager.

✅ Tightening "building envelopes" through improved insulation, more efficient windowpanes, and weatherstripping.

✅ Developing a computerized energy management system to automatically regulate heating and cooling and maintain constant temperature.

✅ Increasing the efficiency of building scheduling for evening, weekends, and holidays, so that heating, cooling, and lighting are needed in a minimum number of areas.

✅ Encouraging students, faculty, and staff to use alternative transportation to campus, including bicycling, walking, and public transportation (see "Transportation," page 64).

✅ Promoting campus community awareness and encouraging participation in energy-conservation programs. Place "Turn off Light" stickers above switches. Establish competitions between residence halls to promote efficiency.

✅ Planting trees strategically to improve the natural cooling of buildings in summer and provide windbreaks in winter.

RESOURCES

Books, Publications, Articles

Carlson, Andrea. *The NIRS Energy Audit Manual: How to Audit Campus, City and other Buildings.* Nuclear Information and Resource Service, 1992. 1424 16th St, N.W., Suite 601, Washington, D.C. 20036; (202) 328-0002.

Cool It! *Students Working for a Sustainable World: The Cool It! Project Directory 1990–91.* Washington, D.C.: National Wildlife Federation, 1991. Contains profiles on campus energy-efficiency projects and organizing tips.

The EarthWorks Group. *30 Simple Energy Things You Can Do To Save the Earth.* Berkeley: EarthWorks Press, 1991.

Energy Task Force. *Energy Management for Colleges and Universities,* 1977. National Association of College and University Business Officers, One Dupont Circle, Suite 510, Washington, D.C. 20036; (202) 861-2500.

Heist, Eric, et al. *Energy Efficiency in Buildings: Progress and Promise.* Washington, D.C.: American Council for an Energy-Efficient Economy, 1986.

Moll, Gary, and Stanley Young. *Growing Greener Cities.* Venice, Calif.: Living Planet Press, 1992.

O'Brien, Kevin, and David Corn. *Energy Conservation: A Campus Guidebook.* Center for Study of Responsive Law, 1981. P.O. Box 19367, Washington, D.C. 20036; (202) 387-8030.

Simpson, Walter, *Recipe for an Effective Campus Energy Conservation Program.* Cambridge, Mass.: Union of Concerned Scientists, 1991.

Snell, Fred, and Walter Simpson. *The University at Buffalo and the Greenhouse Effect.* Buffalo, N.Y.: SUNY Buffalo Press, 1989.

Institutions & Organizations

Alliance to Save Energy, 1725 K Street, N.W., Suite 914, Washington, D.C. 20006; (202) 857-0666.

American Council for an Energy-Efficient Economy (ACEEE), 1001 Connecticut Avenue, N.W., Washington, D.C. 20036; (202) 429-8873.

Association of Physical Plant Administrators of Universities and Colleges, 1446 Duke Street, Alexandria, VA 22314-3492; (703) 684-1446.

Brown is Green, P.O. Box 1941, Brown University, Providence, RI 02912; (401) 863-7837. Request information on energy efficiency.

Conservation and Renewable Energy Inquiry and Referral Service (CAR-EIRS), Box 8900, Silver Spring, MD 20907; 1-800-523-2929

Conserve UB, SUNY Buffalo, Physical Facilities, 120 John Beane Center, Amherst, NY 14260; (716) 636-3636.

Cool It! Project, National Wildlife Federation, 1400 16th St., N.W., Washington, D.C. 20036; (202) 797-5435. Request the energy efficiency information packet.

Ecolympics, Harvard University, Office of Physical Resources, 1746 Cambridge St., Cambridge, MA 02138; (617) 495-3678.

Global Releaf, A Project of the American Forestry Association, P.O. Box 2000, Washington, D.C. 20013; (202) 667-3300. Request information on tree-planting for energy conservation.

Green Lights Program, Environmental Protection Agency, Global Change Division, Office of Air and Radiation, 401 M Street, S.W. (ANR-445), Washington, D.C. 20460; (202) 245-4157.

Lawrence Berkeley Laboratory, Building Science Division, Cycletron Road, Berkeley, CA 94720; (510) 486-4834

Natural Resources Defense Council (NRDC), 40 W. 20th St, New York, NY 10011; (212) 727-2700.

Office of Conservation and Renewable Energy, U.S. Department of Energy, 1000 Independence Avenue S.W., Washington, D.C. 20585; (202) 586-5000.

Rocky Mountain Institute, 1739 Snowmass Creek Road, Snowmass, CO 81654; (303) 927-3128.

The Safe Energy Communication Council (SECC), 1717 Massachusetts Ave, N.W., Suite LL215, Washington, D.C. 20036; (202) 483-8491.

Southern California Gas Company, Marketing Department, 555 W. Fifth Street, Los Angeles, CA 90013; (213) 244-3735.

TreePeople, 12601 Mulholland Drive, Beverly Hills, CA 90210; (818) 753-4600.

Union of Concerned Scientists (UCS), 26 Church Street, Cambridge, MA 02238; (617) 547-5552. Write or call for a complete publications list.

CHAPTER 11: FOOD

While you may not easily detect the connections between an environment at risk and the dinner that was served last night in your campus dining hall, a host of ecological problems can be traced to food choices. The type of meals served, where food comes from, and how it is grown or raised can have an effect on global warming, water pollution, forest destruction, topsoil erosion, as well as how self-sufficient your local farmers may be.

Consider the facts: Producing foods without chemical pesticides benefits the environment by avoiding the water pollution caused by pesticides in runoff and reducing the amount of waste generated by pesticide manufacturing plants. It also protects the health of farmworkers who would otherwise be exposed to toxic chemicals. Buying locally-grown foods rather than products that are shipped from hundreds or thousands of miles away results in less air pollution and fuel consumption due to transportation. Providing vegetarian meals has environmental as well as personal health benefits. Livestock production is a major contributor to topsoil erosion, uses far more water and energy than the production of non-meat foods, and contributes to over half of all water pollution in the United States. Furthermore, over 75 percent of the pesticide residues in the U.S. diet are supplied by meat and dairy products, while only 10 percent are supplied by fruits, vegetables, and grains.

Campus food services can use their resources to foster more sustainable agricultural practices and healthier food choices by purchasing regional and certified organically grown products, by offering a greater selection of meals that are vegetarian (meatless) and vegan (containing no animal products), and by educating the campus community about the connection between diet and the environment.

First, educate yourself. Find out what "certified organic" means, and what is involved in sustainable agriculture. You may also wish to learn more about the environmental effects of raising meat (e.g., water use, pesticide use, etc.).

Assessment Questions

◆ What farm products are grown in your region and when are they available? What processed and pre-made food products (including canned and frozen) are produced locally or regionally, and by what companies? Does your campus buy from local producers and processors? To what extent could food services expand this practice? (See "Research Sources:" B,D,E,F)

◆ Who operates campus food services? Are they run by the college or university, the student association, or outside contractors? (A)

◆ Who is responsible for making decisions about menu-planning and identifying the various food vendors on campus? (A)

◆ Does your school have a food services committee? Do students participate in the committee? (A)

CAMPUS PROFILE:
Hendrix College, Conway, Arkansas

Eating Smart

Motivated by a desire to serve healthy food produced by sustainable agricultural practices, students and administrators at Hendrix College conducted a comprehensive review of campus food resources in 1986. This "Local Foods Project" analyzed the sources, distribution methods, and preparation techniques of the meals served in the school dining hall. The study found that nearly 95 percent of the food served on campus came from out of state, even though agriculture is an important part of the Arkansas economy. Changes in campus food operations resulted in a dramatic increase in the use of locally and organically-produced food sources.

◆ What are your food services' purchasing criteria? Do they consider facts other than cost—such as purchasing organically- and locally-grown produce? (B,F)

◆ Do your food services offer meatless and dairy-free entrees at every meal? Do they purchase any certified organic produce, dairy, or meat products? (C)

◆ Have any surveys been conducted to estimate the demand for vegetarian food? Are any policies or programs promoting vegetarian options in place? If yes, what motivated these policies? (C)

◆ Have food services discontinued the purchase of any food products for environmental reasons? (A)

Research Sources

◆ Your campus food service office should be able to answer questions regarding management of their operations.

B Your food services purchasing manager(s) should be interviewed for food-inventory information. Ask them for access to food-purchasing records.

C The food service director can be contacted about policies relating to organic produce, vegetarian meal-planning, and locally-supplied food. Also inquire about any surveys determining student and staff food preferences.

D Elected officials, local farmers, farm union representatives, and other individuals at the county and state levels may be interested in developing local markets for agricultural commodities. The state departments of agriculture and commerce may be helpful in providing this information.

E Interested faculty members in the agriculture department could be contacted for assistance with your work.

F Hendrix College can provide assistance with your local food-buying project. The publication "Local Food Production for Local Needs: A Manual for the Analysis of a College or University Food Service," provides information on how to conduct food service and local agricultural inventories (see "Resources," page 57).

Recommendations

The studies suggested above will give you a good picture of the food sources in your community and how they find their way to your table. To learn more and to educate others, you and your colleagues may consider:

✔ Conducting a food service inventory to determine types, volume, cost, and origins of food purchased.

✔ Conducting a local agriculture inventory to determine the feasibility of purchasing regionally-produced food.

✔ Conducting a workshop to bring together local growers, food service personnel, and agricultural specialists.

✔ Organizing "health weeks" or a "Great American Meat Out" to feature and promote vegetarian foods and to educate the campus community. Bring experts to campus to discuss how food choices impact the environment.

✔ Purchasing organically-grown and locally-produced foods for campus facilities whenever possible.

✔ Placing flyers or table tents in the dining halls to explain the benefits of organically-grown and local products, and labeling menu items that incorporate these foods.

✔ Offering more vegetarian and vegan menu items in all campus dining halls. The Vegetarian Resource Group provides "Vegetarian Quantity Recipes" for institutional menu-planning.

✔ Placing suggestion boards or boxes in campus dining halls. Contact your student government office and inquire if they have a food service committee. If none exists, create one and use committee meetings to discuss ways to incorporate student suggestions into food buying and menu-planning policies.

✔ Conducting a survey to determine the demand for vegetarian and vegan food and the extent of knowledge on your campus concerning the ecological effects of food production.

✔ Supporting purchases of food products by companies that are ecologically sensitive, such as certified "dolphin safe" tuna and preservative-free packaged foods.

✔ Investigating the possibility of composting food waste for use as mulch or soil amendments. Initiating a food recovery program in which unused food is donated to a local homeless shelter or food bank (see "Solid Waste," page 3).

RESOURCES

Books, Publications, Articles

Campus Favorites: Vegetarian Recipe Collection. Dieticians in College and University Food Service, American Dietetic Association, 605 South Madison St., Lancaster, WI 58313. Contact: Susan Davis Allen.

"A Coast-to-Coast Guide to Vegetarian Organizations." *Vegetarian Times* no. 169, Sept. 1991.

"Education in a Hotter Time: The Campus and the Biosphere in the Twenty-first Century," A Prospectus for the Administrations of Carleton and St. Olaf Colleges for Advancing a Program in Education and Sustainable Development, Aug. 14, 1990. Biology Department, St. Olaf College, Northfield, MN 55057.

Holmes, Hannah. "Eating Low on the Food Chain." *Garbage* Jan./Feb. 1992.

Jacobson, Michael F., Lisa Y. Lefferts, and Anne Witte Garland. *Safe Food: Eating Wisely in a Risky World.* Venice, Calif.: Living Planet Press, 1991.

Klaper, Michael, M.D. *Vegan Nutrition: Pure and Simple.* Umatilla, Fla.: Gentle World, Inc., 1988.

Lefferts, Lisa, and Roger Blobaum. "Eating as if the Earth Mattered." *E Magazine* 1992, 3(1).

Meadowcreek Project. "Local Food Production for Local Needs: A Manual for the Analysis of a College or University Food Service." Fox, AR, Sept. 1988.

"A Proposal to Supply the Hendrix Food Service with Locally Produced Commodities." Office of Student Development and the Hendrix Wellness Project in cooperation with the Meadowcreek Project. Fox, AR, Nov. 1, 1986.

Rifkin, Jeremy. *Beyond Beef: The Rise and Fall of the Cattle Culture.* New York: Dutton, Penguin Books, 1992.

Robbins, John. *Diet for a New America.* Walpole, N.H.: Stillpoint Publishing, 1987.

Valen, Gary, "Hendrix College Local Foods Project." Summary paper. Hendrix College, Conway, AR 72032-3080; (501) 450-1372.

Vegetarian Journal. Vegetarian Resource Group, P. O. Box 1463, Baltimore, MD 21203; (301) 366-VEGE.

Vegetarian Times. For subscription information, call (708) 848-8100.

Werner, Lisa. "Vegetarians on Campus." *Vegetarian Times*, Sept. 1990.

Where Our Food Comes From: The Hendrix College Food System, 1987. A 45-minute video developed by the Hendrix College student research team documenting Hendrix's role in the food system. Available through the Meadowcreek Project.

Where Our Food Comes From: The Oberlin College Food System and Where Our Food Comes From: The Oberlin Student Cooperative Association Food Services, 1988. Reports developed by Oberlin student researchers and Meadowcreek staff. Available through the Meadowcreek Project.

Where Our Food Comes From, 1989. A 30-minute video directed to the campus audience developed by students at Oberlin College in Ohio. Available through the Meadowcreek Project.

Institutions & Organizations

Center for Science in the Public Interest, 1501 16th Street, N.W., Washington, D.C. 20036; (202) 332-9110. Resources available include the Nutrition Action Newsletter, and Safe Food.

Cool It! Project, NWF,1400 16th St., N.W., Washington, D.C. 20036; (202) 797-5435. Request the food chain issue packet.

EarthSave, 706 Fredrick St., Santa Cruz, CA 95062; (408) 423-4069. A nonprofit organization founded by John Robbins that works to educate people about the connection between food choices and environmental and health impacts.

Vegetarian Resource Group, P. O. Box 1463, Baltimore, MD 21203; (301) 366-VEGE. Provides meatless recipes for large numbers of people.

The products your university purchases can have a profound impact on the environment and human health during their production, as they are used, and when they are discarded. Colleges and universities are sizable consumers of high-quality paper products, disposable goods, and a wide variety of chemicals. Purchasing safe and ecologically sound products—such as recycled paper, recycled building materials, items produced without chlorofluorocarbons, and organically-grown produce—is an important step in environmentally sensitive decision making and will also help create markets for environmentally sound products. Your school's choice of paper products is particularly important because of the large quantities used. Our national forests—sources of paper products as well as construction materials—are being decimated at alarming rates. Yet wood is often wasted, and paper constitutes nearly half of the total municipal solid waste stream in the United States. In addition, the purchase of certain tropical hardwoods contributes to the destruction of precious and irreplaceable ecosystems.

A great deal of information is available on sustainable forestry, recycling, and the choice of recycled materials available for purchase. Be prepared to discuss these when you conduct your assessment. Also make sure you understand procurement guidelines, pricing preferences, post-consumer content, "greenwashing," and "green consuming."

Assessment Questions

◆ How are purchasing decisions made on your campus? Is there a central purchasing office? How much autonomy do individual departments have regarding the items they purchase? Do campus entities apart from the main purchasing office—such as the copy shops, computer lab, residence halls, student associations, or medical center—make separate purchasing decisions and/or specific product requests? (See "Research Sources:" A)

◆ Are the student stores and food services part of a larger cooperative buying operation, such as the National Association of College Stores or the National Association of College and University Food Services? (A)

◆ How many reams or tons of high-grade writing and copy paper does your campus purchase annually? What is the cost of this paper? Does your campus buy products made of recycled paper? (Recycled paper products could range from letterhead to salad bowls.) How does their cost compare to the price of virgin-paper products? Are any campus publications or newspapers printed on recycled paper? (B)

◆ Do campus food services use primarily disposable plastic and paper products, washable dishes and utensils, or a combination of both? (A)

◆ Are any tropical hardwoods, such as lauan or mahogany, used in the construction of new buildings or other campus fixtures, such as landscape trellises? Does your school purchase any tropical hardwood furniture? From what kinds of wood is it made? Where is this wood grown? Who supplies it? (A,C)

CAMPUS PROFILE:
University of Illinois, Urbana-Champaign

Buying Recycled

With a campus population of 50,000 and an annual purchase of over $2.5 million for paper products alone, the University of Illinois has tremendous potential to help stimulate markets for recycled products. Recognizing the importance of market development, the university expanded their recycling program in 1989 to include procurement guidelines. The written policy states that "the University will purchase products with recycled material content whenever cost, specifications, standards, and availability are comparable to products without recycled content." In 1991, as a result of this policy, 16 percent of the university's

◆ Does your state have procurement guidelines that mandate or encourage a "pricing preference" for recycled products (paper, construction materials, oil, tires, etc.) purchased by state agencies? A pricing preference means that an agency will pay up to a specified percentage more for a product made from recycled materials than for a similar product made from virgin materials. If so, do these guidelines apply to your university? If not, has your school established purchasing guidelines for recycled products on its own? (A,C)

◆ Is organically-grown produce sold or used in food services? Does your campus have a policy for using energy-efficient lighting and appliances? Are nontoxic cleaning products used anywhere in campus facilities? (A)

◆ To what extent can recycled paper, organic produce, and other environmentally sound products be obtained locally? How do these products differ in cost from the conventional products they would replace? Can suppliers meet your school's volume needs? (C,D)

paper purchases (by sales) contained recycled content. By 1992, this amount had increased to 22 percent.

The policy has primarily guided the purchase of paper products, such as computer and copy paper, envelopes, janitorial supplies, university publications, stationery, food-service products, and miscellaneous forms. However, purchasing managers target other recycled items as well, including floor pads, tires, oil, building materials, reprocessed chemicals, and office supplies. To ensure that the policy is actually influencing procurement decisions, purchasing managers actively identify vendors who can supply products and materials with recycled content. If specific items are not available, the purchasing office will often contact manufacturers directly. As a result, several recycled products not previously available are now supplied, including all of the school's parking tickets! Additionally, the purchasing office has created an extensive tracking process and communicates the program's progress to all purchasing managers.

Product purchases are made on an item-by-item cost basis, and the purchasing department requests separate bids for recycled and virgin-paper products in order to obtain actual costs for recycled products. Although the procurement policy establishes no formal pricing preference, the university has purchased recycled products that cost as much as 15 percent more than their virgin equivalent. However, not all recycled product purchases are more expensive. In fact, some items, such as recycled paper towels, toilet and facial tissues, kitchen wipes, and table covers have cost less, and the recycled computer paper now used on campus is 10 percent cheaper than the product it replaced.

Research Sources

A Administrators should be contacted at your campus purchasing office, in the storehouses, and at various departments for information regarding the type, quantity, and cost of purchases. Also contact purchasing managers for residence halls, medical facilities, and student government. Food services and store managers will have similar information.

B Purchasing managers may be able to tell you about the availability and cost of recycled paper and other environmentally sound alternative products such as janitorial supplies. Contact manufacturers of these products for more information.

C Your state and local government purchasing office can be contacted to find out about procurement policies and their applicability to your school.

D Your library is an excellent resource for materials on "green consuming" and environmentally sound products. Also look for the titles listed in this chapter's "Resources" section.

Recommendations

Discussing the following recommendations with purchasing managers can also sensitize them to the environmental impact of their choices. Here are some discussion ideas:

✔ Purchasing reusable, recyclable, and nontoxic products and those made from recycled materials. These products include nontoxic cleaning supplies, recycled and unbleached janitorial and office supplies, recycled oil, reprocessed chemicals, remanufactured parts, building materials, and food-service products.

✔ Establishing procurement guidelines for recycled paper which give a 5 percent to 10 percent pricing preference. Request recycled paper for departmental uses. Recycled paper should be at least 10 percent post-consumer content.

✔ Encouraging the purchase of nonchlorine-bleached paper products. These include printing, writing, copier and computer paper, letterhead stationery and envelopes, newspapers, magazines, brochures, toilet paper, paper towels, bags of all types, wrapping paper, and coffee filters.

✔ Using recycled newsprint in all campus newspapers and publications. Use white paper in preference to colored paper, which is harder to recycle.

✔ Using reusable dishes and utensils where possible in food services. Where disposables must be used, choose paperware made from recovered fibers. Use unbleached paper products where possible.

✔ Starting a reusable mug program allowing campus patrons to receive a discount on beverages if they bring their own mugs.

✔ Establishing a policy prohibiting the purchase of tropical hardwood products, such as furniture, trellises, and construction materials. Encourage the purchase of domestic and temperate wood alternatives and sustainably harvested rainforest products.

✔ Starting a "Buy Recycled" program in student stores, selling notepads, gift cards, and other items made from recycled materials. Do not recommend the use of "biodegradable" plastic bags (they aren't really biodegradable), and opt for reusable or paper bags (preferably made with recycled fibers).

✔ Purchasing energy-efficient appliances and lighting.

✔ Purchasing copy machines with double-sided capability and fax machines that use bond paper instead of fax paper.

RESOURCES

Books, Publications, Articles

Elkington, John, Julia Hailes, and Joel Makower. *The Green Consumer,* New York: Penguin Books, 1988.

The Greenpeace Guide to Paper. 1017 W. Jackson Blvd, Chicago, IL 60607, 1990. An extensive review of the chemicals used in paper manufacturing.

Hollender, Jeffrey. *How to Make the World a Better Place: A Guide to Doing Good.* New York: William Morrow and Company, Inc., 1990.

National Boycott Newsletter. 6506 28th Avenue, N.E., Seattle, WA 98115; (206) 523-0421.

Recycled Products Guide. American Recycling Market Inc., P.O. Box 577, Ogdensburg, NY 13669; (315) 471-0707; Hotline, 1-800-267-0707. Updated annually.

"Recycling, Recycled Products Procurement, and Waste Reduction." *Campus Administrative Manual.* Urbana-Champaign: University of Illinois, Use of Services and Stores Policies, Section VII/B-9, 1990.

Thompson, Claudia. *Recycled Papers: The Essential Guide,* Cambridge, Mass.: MIT Press, 1992.

Vandenberg, Nancy. "Recycled Materials Procurement." *Resource Recycling.* November/December 1986. Available from the Council on the Environment of New York City, 51 Chambers St., New York, NY 10007; (212) 566-0990.

Wellner, Pam, and Eugene Dickey. *The Wood Users Guide.* San Francisco: Rainforest Action Network, 1991.

Institutions & Organizations

Californians Against Waste, Buy Recycled Program, 926 J Street, Suite 606, Sacramento, CA 95814; (916) 443-5422.

Environmental Defense Fund, 257 Park Avenue South, New York, NY 10010; 1-800-CALL-EDF.

EPA RCRA Hotline for Federal Procurement Guidelines; (703) 941-4452.

Greenpeace, 1017 W. Jackson Blvd, Chicago, IL 60607; (312) 666-3305. Ask for information about their Pulp and Paper Campaign.

Green Seal, P.O. Box 1694, Palo Alto, CA 94302; (415) 327-2200.

Rainforest Action Network, 301 Broadway, Suite A, San Francisco, CA 94133; (415) 398-4404.

Rutgers Recycling Center, Rutgers University, Dudley Road, New Brunswick, NJ 08903; (908) 932-5858.

University of Illinios, Urbana-Champaign, Recycling Coordinator, Physical Plant Services Building, 1701 S. Oak Street, Champaign, IL 61820; (217) 244-7283.

CHAPTER 13: TRANSPORTATION

Transportation policies affect numerous environmental problems, including air pollution, land use for roads and parking lots, traffic congestion, energy consumption, noise, and hazardous waste. Campuses contribute to these problems by operating a fleet of vehicles and attracting the private vehicles of students employees and visitors. Aggressive ride-sharing programs, conversion of fleets to cleaner fuels and improved land-use planning can mitigate these impacts.

Assessment Questions

◆ How many people come to campus each day? (This might be estimated from campus population statistics.) How many vehicles travel to campus daily? (See "Research Sources:" A,F)

◆ How do people get to campus? How do they get around the campus? Try to estimate the percent by each mode (single-occupant vehicle, carpool, vanpool, transit, bicycle, walking, etc). How far do people commute on average? What programs exist to encourage alternatives to the single-occupant vehicle? (A,F,G)

◆ How many total parking spaces are available? How are parking permits distributed and priced? Do carpools get discounts? Does the campus meet the entire parking demand? Is the supply expanding? Who pays for construction of these new spaces, what are the costs, and where are they located? (A)

◆ What transit service is available? Are passes sold on campus? Are they sold at a discount? (A,B,C)

◆ What percentage of traffic on nearby streets comes to and from campus? What are the congestion problems, if any? Has traffic congestion increased or decreased in the surrounding area over the past five years, and by what percentage? (B)

◆ What are the total emissions from private vehicles over a one-year period? This can be calculated from the number of cars, distance traveled, and emissions factors. Use both high and low estimates, if other data are not available, to calculate the average volume of emissions. (B)

◆ Does the area exceed federal ambient air-quality standards? How often? (D)

◆ How many fleet vehicles does the campus operate? Are they owned or leased? Are they serviced and fueled on or off campus? How much waste oil and solvents are produced from the campus garage? How is it disposed of? Is there an oil-recycling program in place? What happens to used car batteries? Solvents, waste oil, and used batteries are often considered hazardous waste. Check federal and state regulations for disposal requirements. (A,E,H)

◆ Does your school own or lease any alternative-fuel vehicles (methanol, propane, natural gas) or electric vehicles? (A)

◆ How many parking lots are located on campus or within two miles of it? How much land use is devoted to motor vehicles and parking? (A)

EDUCATION ENVIRONMENT ENLIGHTENMENT

CAMPUS PROFILE:
University of Illinois at Urbana-Champaign

Reducing Traffic

A recent combined initiative of students, parking services, and the local Metropolitan Transit District has fostered the development of a mass-transit system that serves the University of Illinois. Students voted to increase their transportation fee by $13 per semester, so that they would be entitled to ride the citywide public transportation system for free just by showing their student IDs. Modifications were made to the bus routes to service high-density off-campus student housing areas, and students now take 11,000 rides daily on the metropolitan system, compared to less than 300 daily before. Additionally, the university instituted a campus bus system which runs every five minutes and includes a remote lot shuttle system which runs between a parking lot and the campus. Students now take 15,000 rides daily on this system. According to the Campus Parking and Transportation department, the effect of the changes on the campus parking and traffic situation has been "*very* appreciable."

Research Sources

Ⓐ Campus parking and transportation officials should be interviewed.

Ⓑ City and county transportation or public works departments may be consulted. Regional ride-sharing agencies may also provide carpool and vanpool matching services.

Ⓒ Local transit authorities can provide passenger counts and public transit pass information.

Ⓓ State and local air-pollution control agencies can give you emissions information.

E The RCRA Hotline—1-800-424-9346 or (202) 382-3000 in Washington, D.C.—can answer questions about hazardous wastes from garages.

F The administration may have environmental impact statements for campus development. Campus project planners may have information on traffic, transportation modes, and vehicle emissions.

G Your own survey on commuting patterns could be conducted to obtain data.

H Your campus environmental health and safety office is a useful resource.

Recommendations

Planning for more efficient transportation on campus may involve a number of entities. Take a survey of all concerned departments, then meet with each to discuss the following:

✔ Offering reduced-price and preferential parking for carpools.

✔ Providing matching services and other promotional events to encourage ride-sharing, and helping to form vanpools.

✔ Subsidizing transit (bus and train) passes.

✔ Imposing parking surcharges to discourage driving alone. Revenues can be used for ride-sharing programs.

✔ Converting fleet vehicles to clean fuels (methanol or compressed natural gas, for example).

✔ Purchasing or leasing electric and alternative-fuel vehicles for use in fleet services.

✔ Inspecting and maintaining auto emissions control devices on fleet vehicles.

✔ Making fuel efficiency a high priority in the purchase of new vehicles.

✅ Reducing hazardous wastes and properly disposing of materials from the servicing of university vehicles, recycling waste oil, used batteries, and solvents. Locate the closest recycling service. Also consider purchasing retreaded tires and recycled oil to help develop markets for these products.

✅ Providing bike lanes, bike racks, as well as shower facilities, and lockers for commuters willing to burn off some calories on the way to campus. Ask for these facilities to be included in any forthcoming campus and community development plans.

RESOURCES

Books, Publications, Articles

Brittle, Chris, Natalie McConnell, and Shanna O'Hare. *Traffic Mitigation References Guide.* Washington, D.C.: U.S. Department of Transportation, Metropolitan Transportation Commission, 1984.

California Department of Health Services, Toxic Substances Control Division, Alternative Technology Section. *Hazardous Waste Reduction Checklist Automotive Repair Shops,* October 1988. 4250 Power Inn Road, Sacramento, CA 95826; (916) 324-1807.

Institute of Traffic Engineers. *Traffic Engineering Handbook, Trip Generation and Parking Generation,* Washington, D.C., 1965.

Transportation Research Board. *Transportation Research Record.* Several issues of this periodical have valuable information on ride-sharing. Transportation Research Board, Washington, D.C.; (202) 334-2960.

Institutions & Organizations

Association for Commuter Transportation, 808 17th St., N.W., Suite 2000, Washington, D.C. 20006; (202) 659-0602

Bicycle Federation of America, 1818 R St., N.W., Washington, D.C. 20009; (202) 332-6986.

Cool It! Project, NWF, 1400 16th Street, N.W., Washington, D.C. 20036; (202) 797-5432.

League of American Wheelmen, 6707 Whitestone Road, Suite 209, Baltimore, MD 21207; (301) 944-3399. A national bicycle organization.

CHAPTER 14:
CAMPUS DESIGN AND GROWTH

Designing and planning with the environment in mind from the beginning of the campus construction process or during renovation is one of the best ways to minimize environmental impacts. Inefficient buildings waste both natural and financial resources. Incorporating energy-efficiency criteria into building designs will reduce dependency on fossil fuels and produce considerable savings in utility bills. Many construction materials and methods also impact human health. The Environmental Protection Agency has declared poor indoor air quality a major national health concern. Proper ventilation systems and the use of nontoxic building materials can significantly improve the air we breathe indoors. New construction and renovation also present opportunities for incorporating building materials with recycled content, which helps conserve virgin materials and creates markets for these products. Planning that encourages the use of alternative transportation and protects open space has the added benefits of reducing community environmental problems such as air pollution, noise, traffic, and habitat destruction. By embracing ecologically-based planning, design, and construction principles, our campuses can be models for sustainable growth and development.

This area of inquiry may involve urban planning as well as architecture and business. Familiarize yourself with the nature of environmental impact reports before beginning.

Assessment Questions

◆ What is the history of your college or university? When were the oldest buildings constructed? How large was the original campus? What has its pattern of growth been: gradual or rapid? How much land is now owned or operated by your school and what is its geographical dispersement? (See "Research Sources:" A,B,D,E)

◆ What are your school's current plans for expansion and renovation? Does your school have a long-range development plan describing existing and future land uses for the campus? Is your campus currently constructing new facilities or does it have plans for development in the near future? (B,C)

◆ Does your college or university have an ongoing planning committee? Do the members include proportionate members of administrators, faculty, students and community representatives? (B,C)

◆ If your university has an architecture and/or planning school, have students and faculty from these departments been involved in the campus growth planning and building design process? (B,C)

◆ Does your school's planning document or long-range development plan contain environmental criteria? Do the goals of the plan discuss "preservation and enhancement of historic buildings and open space," or other related goals for the constructed and natural environment? (B)

CAMPUS PROFILE:
World College West, Petaluma, California

One Smart Campus

Since their completion in 1985, all of the permanent academic buildings and residence halls at World College West employ passive solar heating and natural lighting. The buildings are not air-conditioned, and overhead lights are only used on dark days. Campus toilets use one-half gallon of water per flush, and the campus has installed energy-efficient halogen lights outdoors. Though 40 percent of the students and all faculty commute, the campus itself is pedestrian, and all drivers must park at a remote parking lot and take a shuttle bus to the campus. Buildings and roads cover only 10 percent of the 194-acre campus, while the rest of the grounds are covered with native landscaping where deer, fox, and other wildlife are free to roam.

◆ Do environmental impact statements for new projects consider the cumulative environmental impacts of university expansion on traffic, pollution, etc.? What strategies for mitigating these impacts are recommended? (B,H)

◆ Do campus planning documents contain statements about coordinated planning efforts between your college and the town or city in which it is located? Is your school exempt from any local (city or county) land-use planning and zoning laws? (B,H)

◆ Are local neighborhood groups involved in monitoring campus expansion issues? Have these associations challenged any new growth on campus? What are the most significant, visible environmental impacts of the campus on the surrounding community? What impact does the surrounding community have on your campus, especially if it is in an urban area? (D)

◆ Are there plans to demolish any buildings which may be considered historic? What will be built in their place? Are there any building-restoration projects? (B,E)

◆   Does your school own any land that is being developed for private, noneducational facilities or is such development planned for the future—as, for example, an office park, shopping center, non-staff or student housing? (B,C,H)

◆   Are there examples of environmentally sound building design on campus, such as those incorporating passive solar technology, energy efficiency technologies, nontoxic or recycled building materials? (B,C,F,G)

◆   Does your campus have cooperative housing that promotes the application of appropriate technology and environmentally sensitive lifestyles? How could such housing be established? (C,I)

◆   What are the building codes in your community that affect building on campus? Do they include environmental criteria—such as building materials and energy use? (B,H)

Research Sources

Ⓐ   The library and campus catalog will have historical background information about your university's growth over the years.

Ⓑ   The capital programs or planning office will have your school's long-range development plan or related land-use planning documents. The office should also have copies of environmental impact statements (EISs) for campus expansion projects. Also contact the planning office regarding campus renovation plans.

Ⓒ   Your school's planning committee is an excellent resource. Interview members and attend meetings. Also talk with faculty members who teach architecture, urban planning, and natural resources management.

D Neighborhood associations and local environmental groups can also be contacted, if they have been involved with college- or community-planning issues. Also contact your town or city planning office for information.

E The National Register of historic buildings in Washington, D.C. may include buildings on your campus. Often, entire college campuses are registered as historic districts. If there is a proposal to tear down such buildings, and the federal government is involved in any way (e.g., if the government will fund the demolition, or is involved in the building that is being erected on the site) then federal law requires a Section 106 Review. This is a public hearing about the "environmental impact" of the alteration. Check with your local historical society for more information.

F The manufacturers and suppliers of various building materials and products can be contacted for information regarding product extraction, harvesting, or chemical composition. Request material safety data sheets (MSDSs) for products and information about recycled content.

G Your campus or community library will also have information on environmentally sound building products and systems, and indoor air-quality issues. Look for "Resources" listed in this section and related sections in this book, such as "The Workplace Environment," "Transportation," "Energy," and "Procurement Policies."

H Your county or city planning office maintains building codes and may also provide impact reports.

I Your Residential Hall Association, Student Government, and Commuters Association can provide information about co-ops. If they exist, talk to students who live there.

Recommendations

Meet with administrators and student government representatives involved in campus planning, building, and housing—as well as with other interested community members and political representatives—to discuss ways to implement the following:

✔ Reviewing the environmental and health impacts of building construction. Consider the health hazards that may be posed by carpets, paints, compressed wood products, and other building materials. Also consider the environmental impacts of product extraction, production, and disposal.

✔ Incorporating energy efficiency into building design and renovation. Systems, technologies, and products include solar power, daylighting, compact fluorescent bulbs, room occupancy sensors, energy efficient appliances, and computerized energy-management systems.

✔ Using renewable and recycled materials in building construction. Examples include recycled carpet padding, "good wood" (wood not harvested from rainforests or ancient forests), recycled steel, "glassphalt" (asphalt made with recycled glass), and insulation made with recycled paper.

✔ Using nontoxic building materials and supplies wherever possible. Examples include nonsynthetic carpets, nontoxic paints and adhesives.

✅ Making use of the long-range development planning process to integrate environmental concerns into campus design. Encourage growth, renovation, and mitigation measures which are compatible with strong environmental standards. Promote student, faculty, and staff participation in the public hearing process for review of campus EISs.

✅ Establishing student housing cooperatives. If your school doesn't already have one, recommend establishing student housing cooperatives that serve as models of sustainable living. Buildings can incorporate solar collectors, passive solar architecture, energy-efficient appliances and lighting, and water-reclamation systems. Residents can participate in recycling, composting, gardening, gray water systems, and "green consuming."

✅ Preserving and enhancing green space. Support tree-planting programs and cooperative campus gardens. Landscape with native species and, if your school is in a region where drought is a concern, use drought-tolerant species. Trees and green spaces act as buffers against noise, provide shade and wildlife habitats, protect against erosion, reduce city temperatures, reduce building energy consumption, add natural beauty, and act as a natural control against global warming.

✅ Analyzing the environmental impacts of campus expansion projects. Specifically, review the potential impact on traffic congestion, wildlife habitats and historic buildings, noise and air pollution.

RESOURCES

Books, Publications, Articles

Canfield, Chris, ed. *Conference Report: The First International Ecological Cities Conference 1990.* Conference held at the University of California, Berkeley. Berkeley: Urban Ecology and Cerro Gordo Town Forum, 1990.

EARTHWORD. The quarterly journal of the EOS Institute, 1550 Bayside Drive, Corona Del Mar, CA 92625; (714) 644-7111. Primary sponsors of EOS are Architects, Designers and Planners for Social Responsibility and the Permaculture Institute of Southern California.

First Los Angeles Ecological Cities Conference. *Resource Guide.* Typescript. Los Angeles: UCLA, 1990. Available through the Eco-Home Network, 4344 Russell Ave., Los Angeles, CA 90027. Contains expanded list of resources on ecological design and development.

Lechner, Norbert, *Heating, Cooling, Lighting: Design Methods for Architects,* John Wiley & Sons, Inc., 1991. 605 3rd Street, New York, NY 10158.

Mollison, Bill. *A Practical Guide for a Sustainable Future.* Washington D.C.: Island Press, 1990.

Our Common Future: Report from the World Commission on Environment and Development, Oxford University Press, 1989.

Recycled Products Guide. American Recycling Market, Inc., P.O. Box 577, Ogdenburg, NY 13669; (315) 471-0707. Contains information on recycled building materials. Published annually.

Register, Richard. *Eco-City Berkeley: Building Cities for a Healthy Future*. Berkeley: North Atlantic Books, 1987.

U.S. Environmental Protection Agency, Office of Air and Radiation. *The Inside Story: A Guide to Indoor Air Quality*, Sept. 1988, Report EPA/400/1-88/004. U.S. EPA, Office of Air and Radiation, Washington, D.C. 20460.

Van der Ryn, Sim, and Peter Calthrope. *Sustainable Communities: A New Design Synthesis for Cities, Suburbs and Towns*. San Francisco: Sierra Club Books, 1986.

Wellner, Pamela, and Eugene Dickey. *The Wood Users Guide*. San Francisco: Rainforest Action Network, 1991.

Institutions & Organizations

American Community Gardening Association, Philadelphia Green, 325 Walnut Street, Philadelphia, PA 19106.

American Institute of Architects, 1350 New York Avenue, N.W., Washington, D.C. 20006; (202) 626-7300. Request the guide on environmentally sound building materials put together by the AIA's Environmental Resources Committee.

Architects/Designers/Planners for Social Responsibility (ADPSR), 225 Lafayette Street, New York, NY 10012; (212) 431-3756.

Center for Resourceful Building Technology, P.O. Box 3413, Missoula, MT 59806; (406) 549-7678.

Cool It!, a project of the National Wildlife Federation, 1400 16th Street, N.W., Washington, D.C. 20036; (202) 797-5435. Request the tree-planting issue packet.

Eco-Home Network, 4344 Russell Avenue, Los Angeles, CA 90027; (213) 662-5207.

Energy Efficient Builders Association (EEBA), Technology Center, University of South Maine, Graham, ME 04038; (207) 780-544.

E² (Environmental Enterprises), Environmental Building Consultants, 12915 Greene Avenue, Los Angeles, CA 90066; (310) 827-1217.

Global ReLeaf, American Forestry Association, P.O. Box 2000, Washington, D.C. 20013; (202) 667-3300

Institute for Regenerative Studies, California Polytechnic University, 3801 W. Temple Avenue, Pomona, CA 91768; (714) 869-2684. Plans include a 16-acre student community to explore new systems for energy, water, shelter, food production, and waste disposal.

North American Students of Cooperation (NASCO), Box 7715, Ann Arbor, MI 48107; (313) 663-0889.

Passive Solar Industries Council, 109 Vermont Avenue, N.W., Suite 1200, Washington, D.C. 20005; (202) 371-0357.

Permaculture Institute of Southern California, 1027 Summit Way, Laguna Beach, CA 92651; (714) 494-5843.

Rainforest Alliance, 270 Lafayette Street, Suite 512, New York, NY 10012; (212) 941-1990. Ask for information on their "Smart Wood" program.

Rocky Mountain Institute, 1739 Snowmass Creek Road, Snowmass, CO 81654; (303) 927-3128.

Safe Buildings Alliance, Metropolitan Square, 655 15th Street, N.W., Suite 12, Washington, D.C. 20005; (202) 879-5120.

Sustainable Enterprises Explorations, HUD-Room 724, Washington, D.C. 20410; (202) 708-2504.

TreePeople, 12601 Mulholland Dr., Beverly Hills, CA 90210; (818) 753-4600.

United States Cooperative Association, 2424 Ridge Rd., Berkeley, CA 94704; (510) 848-1936.

Urban Ecology, P.O. Box 10144, Berkeley, CA 94709; (510) 549-1724

SECTION III:
THE BUSINESS
OF EDUCATION

THE COLLEGE OR UNIVERSITY OF THE
1990s does not exist in an economic vacuum. Although its primary purpose is to educate, the modern-day institution of higher education is also a business—and a big one. The average college or university has a sizeable portfolio of investments in corporations on every continent. It owns real estate and other tangible assets. Government and corporate research grants are also significant factors in the university's balance sheet.

SECTION III: THE BUSINESS OF EDUCATION

And like any business, the university has myriad relationships with vendors, suppliers, and consumers. Many of these business relationships have ramifications for environmental quality, both on and off campus. Understanding these complex relationships and their effect on the wider environment is a challenging task. Playing an active role in the university's business decision-making is an achievable goal for motivated students.

Much university research produces valuable advances in health and social welfare. However, as funding sources become more and more scarce, universities are increasingly relying on private corporations and large government agencies for research funding, and therefore producing more of the kind of research these sponsors want to see—including pesticide research funded by chemical corporations, nuclear power research funded by the Department of Energy, or weapons development supported by the Department of Defense. Continuing emphasis on developing profitable research can often supersede the environmental consequences of that research. However, careful scrutiny of university research objectives and funding sources can provide the basis for steering research activities in a more ethical and environmentally beneficial direction.

Assessment Questions

◆ Do you attend a "college" engaged in research? Universities are not the only schools that act as research institutions; as many as 50 four-year liberal arts colleges in the United States have also been classified as "research colleges." (See "Research Sources:" A)

◆ What are the largest research projects at your school in terms of funding? What are the titles of the research projects, the affiliated campus departments and researchers? What is the stated purpose of these projects? (A,B)

◆ How much research funding does your school receive? What are the funding sources? These could include the federal government (e.g., Department of Health Services, National Science Foundation, Department of Energy, Department of Defense), state, county or city agencies, foundations and charitable trusts or private corporations. How does the total amount of research grants and contracts at your school compare with other universities? (A,B)

◆ Who are the clients of the research? Do they differ from the funding sources? (A,B)

◆ Is any environmental-protection research conducted on campus? This might include research projects related to toxics or hazardous-waste-management, recycling, pollution prevention, renewable energy, integrated pest management, or alternative agriculture. (A,B)

◆ What are the specific subjects of this environmental research, if any? How much of this research is related to preventing pollutants or wastes from being formed or preventing resource depletion, and how much focuses on controlling pollution or disposing of wastes after they are formed? (B)

◆ How much funding is available for environmental research? What is the funding source? Which department is the recipient of those funds? (A,B)

◆ What is the environmental impact of the research conducted at your college or university? Could any projects be considered environmentally destructive or hazardous? These might involve pesticide development, nonsustainable agriculture, resource-extraction techniques, or the disposal of hazardous wastes. (A,B)

◆ Does your school conduct military research? What is the nature of this research? Is your university conducting classified research? (A,C)

◆ Is your school part of a cooperative university-state government research and policy center? If so, what campus representatives sit on the steering committee? What kinds of research projects have been conducted? (A)

◆ Do any of the faculty conducting research at your school consult and/or conduct research at outside corporations that have a record of environmental violations or are engaged in activities—such as deforestation or toxic-chemical manufacturing—that are known to harm the environment? (B,D,E)

Research Sources

Ⓐ Your school's office of contracts and grants can provide annual reports. These will provide information on the number of awards to your school, their dollar value, funding agencies, and departmental recipients. Also ask for separate abstracts of your school's listings for government- and corporate-sponsored research.

Ⓑ Your school libraries, research departments, and laboratories may be contacted directly for annual reports and newsletters relating to specific research programs.

Ⓒ The Department of Defense (DOD) will provide "Work Unit Summary Sheets" when you ask for them through a Freedom of Information Act (FOIA) request directed to the Defense Technical Information Center. Request all sheets relevant to the research conducted at your school.

Ⓓ The office of contracts and grants, or the chancellor or president's office can tell you how to obtain a copy of your school's "guidelines" regarding university research and a listing of faculty and administrators who serve on administrative committees responsible for overseeing government- and corporate-sponsored research.

Ⓔ The Department of Defense, University Presidential and Federal Advisory Committees will have membership listings and reports. Look for professors and administrators from your school who are on these committees to serve as sources of information.

Recommendations

Your meetings here may be very informative to the larger community. Consider a public forum to develop a dialogue concerning the following:

✔ Encouraging research that has environmentally beneficial objectives, such as pollution prevention, toxic-use reduction, recycling technologies, renewable energy, xeriscaping (drought-resistant landscaping), water-conservation technologies, and energy efficiency.

✔ Discouraging research that promotes environmentally destructive activities, such as research into broad-spectrum pesticides or nuclear power. Promote pollution prevention as well as pollution control and clean-up strategies.

EDUCATION ENVIRONMENT ENLIGHTENMENT

CAMPUS PROFILE:
University of California

Looking Toward the Future

The University of California Statewide Integrated Pest Management (IPM) Project is a prime example of environmental research that has paid off. The IPM Project has established itself as the national leader in developing economical and environmentally sound techniques for controlling pests in agriculture. In its first decade, the research program funded over 200 projects in 35 different commodities. Drawing the the expertise of hundreds of scientists in a variety of disciplines, the project's research has provided growers with practical techniques for reducing their reliance on synthetic pesticides. These strategies include using biological and cultural methods, such as introducing natural predators to control pests. The UC's IPM Project has resulted in significant reductions in synthetic pesticide applications. For instance, in one program, growers were able to reduce their aphid-control sprays for Brussels sprouts from seven applications to one. And, when California almond growers adopted IPM strategies in the early 1980s, their insecticide use dropped by 45 percent.

The education arm of the project ensures that IPM techniques are being implemented in the field. An extensive computer database, numerous IPM manuals and publications, and annual training conferences for pest management professionals have helped California growers adopt environmentally based pest control strategies, thereby reducing the economic, environmental, and health risks associated with synthetic pesticides.

✅ Reducing or eliminating research into weaponry, biological and chemical warfare, and other military technologies.

✅ Establishing an advisory committee for oversight of research conducted at the university, if none already exists. A broad cross-section of the campus should be represented, including students.

✅ Setting up a board consisting of faculty, students, and community members for the specific purpose of targeting sources of funding for environmentally beneficial research.

RESOURCES

Books, Publications, Articles

It's Only the Beginning. Newsletter of People for a Socially Responsible University, 358 N. Pleasant St., Suite 307, Amherst, MA 01002.

McMillen, Liz. "Quest for Profits May Damage Basic Values of Universities, Harvard's Bok Warns." *The Chronicle of Higher Education,* April 24, 1991.

Noble, David F. "Higher Ed Takes the Low Road: In a battle between education and profit-oriented research, guess what wins?" *Newsday,* Oct. 8, 1989.

Noble, David F. "Tuition and the Education Business." *USA Today,* Jan. 1991.

"Universities and the Military." *The Annals of the American Academy of Political and Social Science,* vol. 502, Philadelphia, 1989.

"When the Price is Too High: Colleges need funds but they fear selling out." *Newsweek on Campus.* November 1987.

Institutions & Organizations

Foundation for Economic Trends, 1130 Seventeenth Street, N.W., Suite 630, Washington, D.C. 20036; (202) 446-2823. For information on biological warfare research at universities.

People for a Socially Responsible University, 358 N. Pleasant St., Suite 307, Amherst, MA 01002; (413) 549-4625.

SEAC, P.O. Box 1168, Chapel Hill, NC 27514; (919) 967-4600. Request information on campuses and military research.

War Research Information Service, P.O. Box 748, Cambridge, MA 02142; (617) 354-9363.

The size of most colleges and universities allows them to have tremendous economic leverage, which can be used to extend the ethic of environmental responsibility to campus suppliers and vendors. It is important for those concerned with ecological issues to investigate not only the environmental impacts of the products your school purchases but also the environmental practices of the companies selling products and contracting for campus services. This is economic ecology and, depending on the volume of your school's purchases, can have wide-ranging consequences.

Assessment Questions

◆ Does your school have a policy that makes the environmental practices of a company a criterion in selecting campus suppliers and vendors? (See "Research Sources:" C)

◆ Has your school chosen to stop doing business with a particular company because of environmental violations (or because of other reasons, such as connections with weapons manufacturing)? (C)

◆ What are the environmental records of companies that have large contracts with your school? Since it may not be possible to investigate all vendors, you may wish to target selected businesses. You may also wish to investigate social and labor practices and connections with weapons research and manufacturing. Analyze the environmental record of a vendor's parent companies as well. (A,D)

◆ What do your campus stores buy and from whom? Ask your student store managers to provide a list of products sold, their manufacturers, and parent companies. (C)

◆ Does your school contract for management of major services such as food service or student store operations? Investigate the environmental records of such contractors. (C)

◆ Are any companies with which your school does business involved in nuclear power, clear-cutting of ancient forests or rainforests, or have any been identified as polluters? Which ones? (A,D)

◆ Have any of your school's vendors, suppliers, contractors, or services signed the Valdez Principles? Which ones? (B)

THE VALDEZ PRINCIPLES

In 1989 the Coalition for Environmentally Responsible Economies (CERES) developed a set of 10 principles for corporate environmental responsibility, called the "Valdez Principles." These principles are designed to commit businesses to protecting the environment through their actions and policies and are one way of evaluating university and corporate responsibility.

Introduction

By adopting these Principles, we publicly affirm our belief that corporations have a responsibility for the environment, and must conduct all aspects of their business as responsible stewards of the environment by operating in a manner that protects the Earth. We believe that corporations must not compromise the ability of future generations to sustain themselves.

We will update our practices continually in light of advances in technology and new understandings in health and environmental science. In collaboration with CERES, we will promote a dynamic process to ensure that the Principles are interpreted in a way that accommodates changing technologies and environmental realities. We intend to make consistent, measurable progress in implementing these Principles and to apply them in all aspects of our operations throughout the world.

The Valdez Principles

❶ Protection of the Biosphere
We will reduce and make continual progress toward eliminating the release of any substance that may cause environmental damage to the air, water, or the earth or its inhabitants. We will safeguard all habitats affected by our operations and will protect open spaces and wilderness, while preserving biodiversity.

❷ Sustainable Use of Natural Resources
We will make sustainable use of renewable natural resources such as water, soils, and forests. We will conserve nonrenewable natural resources through efficient use and careful planning.

③ Reduction and Disposal of Waste

We will reduce and where possible eliminate waste through source reduction and recycling. All waste will be handled and disposed of through safe and responsible methods.

④ Wise Use of Energy

We will conserve energy and improve the energy efficiency of our internal operations and of the goods and services we sell. We will make every effort to use environmentally safe and sustainable energy sources.

⑤ Risk Reduction

We will strive to minimize the environmental, health, and safety risks to our employees and the communities in which we operate through safe technologies, facilities, and operating procedures, and by being prepared for emergencies.

⑥ Marketing of Safe Products and Services

We will reduce and where possible eliminate the use, manufacture, or sale of products and services that cause environmental damage or health or safety hazards. We will inform our customers of the environmental impacts of our products or services and try to correct unsafe use.

⑦ Environmental Restoration

We will promptly and responsibly correct conditions we have caused that endanger health, safety, or the environment. To the extent feasible, we will redress injuries we have caused to persons or damage we have caused to the environment and will restore the environment.

⑧ Informing the Public

We will inform in a timely manner everyone who may be affected by conditions caused by our company that might endanger health, safety, or the environment. We will regularly seek advice and counsel through dialogue with persons in communities near our facilities. We will not take any action against employees for reporting dangerous incidents or conditions to management or to appropriate authorities.

⑨ Management Commitment

We will implement these Principles and sustain a process that ensures that the Board of Directors and Chief Executive Officer are fully informed about pertinent environmental issues and are fully responsible for environmental policy. In selecting our Board of Directors, we will consider demonstrated environmental commitment as a factor.

⑩ Audits and Reports

We will conduct an annual self-evaluation of our progress in implementing these Principles. We will support the timely creation of generally accepted environmental audit procedures. We will annually complete the CERES Report, which will be made available to the public.

Research Sources

◆ National environmental groups and their publications can provide information on companies with poor environmental records, such as Environmental Action's annual listing of the "Filthy Five" corporate polluters.

Ⓑ The Coalition of Environmentally Responsible Economies (CERES) can provide a list of companies that have signed or are in the process of signing onto the Valdez Principles.

Ⓒ Your school's purchasing or business administration office can be contacted for information on procurement criteria and for a list of companies with whom your school does business.

Ⓓ Individual companies may be contacted directly for information regarding their environmental practices.

Recommendations

Those who make the business decisions on your campus need to be identified first. When meeting with these individuals, offer to participate in implementing the following:

✔ Establishing an environmental responsibility policy allowing the campus community to scrutinize the practices of companies with whom your school does business. A committee should be established to review those companies violating the policy. This policy should include recognition for companies with excellent environmental records.

✔ Developing environmental criteria for campus suppliers and vendors. Examples include recommending that suppliers use recycled packaging, use recycled paper for correspondence, and refrain from using styrofoam "peanuts" or nonrecyclable plastic-window envelopes. Suppliers will need to be informed about these campus policies.

✔ Encouraging vendors to adopt the Valdez Principles. Your Student Association and University Board of Trustees can adopt these principles as well.

✔ Insisting that your school cease doing business with environmentally irresponsible companies until reforms have been made. To accomplish this objective, you can organize a petition drive and/or call a meeting with the appropriate campus administrators. Presenting a list of alternative products and companies will help your efforts.

✔ Educating the campus community about avoiding certain targeted products in student stores, and offering alternatives. Explain why you are asking patrons to select alternative products. List the alternatives that your store offers.

RESOURCES

Books, Publications, Articles

Building Economic Alternatives. Quarterly of Co-Op America. 2100 M Street, N.W., Suite 403, Washington, D.C. 20063; (202) 872-5307.

CAMPUS PROFILE:
University of California, Los Angeles

Ethical Questions

In 1987, students on the Board of Directors of the Associated Students UCLA (ASUCLA) developed a policy that allows anyone from the campus community to scrutinize the companies with whom the association does business. ASUCLA, one of the largest associated student operations in the country with an annual budget of over $80 million, primarily manages food services, student stores, and other student services on campus. As a result of the policy, the Board of Directors in 1989 discontinued buying General Electric products because of the company's environmental violations, as well as labor practices and connections with weapons manufacturing.

The CERES Coalition. *The 1990 CERES Guide to The Valdez Principles.* 711 Atlantic Avenue, Boston, MA 02111; (617) 451-0927.

The Council on Economic Priorities. *Shopping for a Better World.* 30 Irving Place, New York, NY 10003; (212) 420-1133.

The EarthWorks Group. *50 Simple Things Your Business Can Do to Save the Earth.* Berkeley: EarthWorks Press, 1989.

The Green Consumer Letter. Tilden Press, Inc., 1526 Connecticut Ave., N.W., Washington, D.C. 20036; 1-800-955-GREEN. Monthly newsletter with information on environmentally sound products.

Hollender, Jeffrey. *How to Make the World a Better Place: A Guide to Doing Good.* New York: William Morrow and Company, Inc., 1990.

National Boycott Newsletter. 6506 28th Ave., Seattle, WA 98115; (206) 523-0421.

Institutions & Organizations

ASUCLA, Office of the Executive Director, 308 Westwood Plaza, Kerckhoff Hall, University of California, Los Angeles, CA 90024; (310) 206-8011. For information about the Policy on Ethical and Social Responsibility, request the brochure, "Let Your Conscience Be Our Guide."

Co-Op America, 2100 M Street, N.W., Suite 403, Washington, D.C. 20063; 1-800-424-COOP.

The Council on Economic Priorities, 30 Irving Place, New York, NY 10003; (212) 420-1133.

Environmental Action, 1525 New Hampshire Avenue, N.W., Washington, D.C. 20036; (202) 745-4870.

Infact, 256 Hanover Street, Third Floor, Boston, MA 02113; (617) 742-4583.

olleges and universities collectively hold billions of dollars in corporate investments and, therefore, have tremendous fiscal clout. Through these portfolios, this nation's educational institutions can promote practices that are polluting and destroying the environment, or they can reinforce environmentally sound business practices. Establishing environmental criteria for investments also makes good business sense. Financial managers confirm that ethical investments made with the environment in mind can have positive financial results, often performing better than traditional stock and bond options. In the 1980s we witnessed the success of the movement to bring about divestment from companies doing business in South Africa. Students in the '90s can use the same approach to benefit the environment.

Assessment Questions

◆ What is your school's current investment portfolio? Look for the company names, the number of shares of stocks or bonds held in each company, their "book" (original) value, and their current value. (See "Research Sources:" A,B)

◆ What is the total amount of your school or school system's investment holdings? Does your school rank among the largest investors in any particular company? (A)

◆ What campus entities maintain their own investment portfolios? (E)

◆ Who is responsible for managing campus investments? An independent financial management firm or an office of the college itself? (A)

◆ Is your campus part of a state school system? Who has the authority to change the investment policy? (Usually this is the board of trustees or regents.) (A)

◆ Is there a special advisory committee that makes recommendations regarding investments, or does this charge fall under the responsibilities of the board of trustees' finance committee? (A)

◆ Does any part of the institution's investment policy or guidelines discuss environmental screenings? Do these guidelines discuss other social criteria? These could include investments related to South Africa, human rights violations, occupational safety and health issues, fair employment practices, weapons production, consumer protection, animal testing, or "sin stocks" (alcohol, tobacco, and gambling). Is your school violating any of its investment guidelines? (A,B,F)

◆ Does your school have holdings in companies involved in clear-cutting of American old-growth forests or tropical rainforests? What about development of nuclear energy;

illegal dumping of hazardous wastes? In what specific destructive actions are these corporations involved? (B,C)

◆ Are any members of the college board of trustees or regents also members of the boards of directors for companies with whom your school holds investments? Are any board of trustee members large shareholders of companies in which your school invests? (A,D)

Research Sources

Ⓐ The school's business office or treasurer should have a copy of your school's investment portfolio, investment policy, and investment guidelines. If your school is part of a state school system, you will most likely need to contact the main business office for university administration. If campus officials deny you access to portfolio information, there are several strategies you can use to obtain the materials you are looking for. If you attend a public school, you can request the information by using the Freedom of Information Act (see "Research Tips," page xxii). At a private school where administrators do not have to release the portfolio, you can apply pressure by using campus media and getting support from faculty and alumni members.

Ⓑ Economics professors, business students, or representatives from investment firms can provide assistance in interpreting portfolio information. Individuals from "socially responsible investment" (SRI) companies may provide free services.

Ⓒ Environmental groups or SRI firms that monitor corporate environmental practices can provide listings of environmentally sound or irresponsible companies. Resources from The Council on Economic Priorities provide a good start for developing ethical investment criteria.

Ⓓ Your library may have company annual reports and 10K retirement or pension fund reports. If not, call the company's public information office and request that they be mailed to you. Review shareholder resolutions listed in these reports. This will give you a good indication of the types of practices progressive shareholders are trying to reform.

Ⓔ Contact the Student and Alumni Associations on campus, for example.

Ⓕ Other schools with socially responsible investment policies or practices such as Hampshire College, Williams College, and Harvard University may also be good resources.

Recommendations

An environmental investment advisory committee should be established, comprised of students, faculty, administration, the college treasurer, alumni, and trustees. This advisory committee can be charged with providing guidance on investment decisions and establishing environmental investment criteria. The focus of this group should be:

✔ Developing an environmentally responsible investment policy. The goal is to reform current investment policy to reflect environmental considerations. Some companies may have an excellent track record in one area yet poor in another. Selecting which companies to invest in or divest from requires extensive research and depends on the priorities of the institution and the process of evaluating a company's products and business practices. You may wish to eventually expand the investment policy to include other social and ethical criteria.

CAMPUS PROFILE:
Hampshire College
Amherst, Massachusetts

Investing in the Environment

Student activists at Hampshire College have been working on the issue of socially responsible investment since the late 1970s, prompting the college to develop a comprehensive socially responsible investment policy and become the first school to divest from companies in South Africa. In 1991 students involved with People for a Socially Responsible University (PSRU) and the Hampshire Reinvestment Coalition (HaRC) used the policy to pressure the board of trustees to divest from corporations engaged in major U.S. Department of Defense contracting.

The Hampshire College Investment Policy embraces the objective of optimizing financial return while adhering to stringent ethical investment principles. The policy, which has been strengthened since its adoption in 1977, establishes an advisory committee called the Committee on Investment Responsibility (CHOIR). The committee, comprised of trustees, faculty members, students, alumni, and the treasurer, is responsible for making shareholder recommendations to the full Board of Trustees and for monitoring investment activities. Environmental criteria are an important part of the framework used for investment decisions. The policy's guidelines, for example, favor holdings in companies that provide a safe and healthy work environment and that demonstrate innovation in environmental protection, with regard to product development, adherence to regulations, pollution prevention, and waste reduction.

✔ Networking with people from the other campuses in developing an environmentally responsible investment movement, especially if your school is part of a state-wide campus system.

Ethical Investment Strategies

In their book Ethical Investing, Amy Domini and Peter Kinder summarize the three approaches to ethical investing. All three strategies complement one another. The most comprehensive socially responsible investment guidelines will contain all three elements.

1) The Avoidance Approach: Refusing to invest in companies whose products, services, or business practices are not supported by your institution. The South Africa divestment movement is one of the best-known examples of the avoidance strategy.

✔ Divest from companies that corporate monitoring organizations have targeted as environmentally irresponsible, such as those on Environmental Action's annual "Filthy Five" corporate polluter list, companies targeted by the EPA as serious polluters or repeat violators of environmental standards, or the U.S. Department of Defense's Top 100 Contracts list.

2) The Positive Approach: Actively buying stocks and bonds in companies whose products, services, and business practices demonstrate respect for the environment or other social goals.

✔ Encourage investments in companies that have signed the Valdez Principles.

✔ Encourage investments in companies involved in renewable and safe energy development, sustainable agriculture and forestry, environmentally sound waste-reduction techniques, manufacturing of products made from recycled materials, and other environmentally beneficial activities.

3) The Activist Approach: Investing in objectionable companies whose practices your institution wants to change and using shareholder rights and resolutions to educate and influence corporate policy. Although this approach may seem merely symbolic, many corporations do take proxy votes seriously, initiating positive reforms as a result.

✔ Introduce proxy resolutions and use university proxy votes to support resolutions designed to create environmental reforms.

✔ Encourage corporations to adopt the Valdez Principles.

RESOURCES

Books, Articles and Publications

Alperson, Myra, et al. and The Council on Economic Priorities. *The Better World Investment Guide: One Hundred Companies Whose Policies You Should Know About Before You Invest Your Money.* Englewood Cliffs, N.J.: Prentice-Hall, 1991. Contains additional resources.

The CERES Coalition. *The 1990 CERES Guide to The Valdez Principles.* 711 Atlantic Avenue, Boston, MA 02111; (617) 451-0927.

Domini, Amy L., and Peter D. Kinder. *Ethical Investing: How to Make Profitable Investments Without Sacrificing Your Principles*. Reading, Mass: Addison-Wesley Publishing Company, Inc., 1986.

Franklin Research and Development. *Asking Ethical Questions*. Boston, Mass., 1992; (617) 423-6655. A useful guide to rating companies.

Meeker-Lowry, Susan. *Economics As If the Earth Really Mattered: A Catalyst Guide to Socially Conscious Investing*. Philadelphia: New Society Publishers, 1988.

People for a Socially Responsible University. "Investment/Divestment Project." Amherst, Mass. 1990. Organizing materials for a socially responsible campus investment campaign.

Simon, John, et al. *The Ethical Investor: Universities and Corporate Responsibility*. New Haven: Yale University Press, 1972.

"A Socially Responsible Financial Planning Guide." A publication of Co-Op America, 2100 M Street, N.W., Suite 403, Washington, D.C. 20063; (202) 872-5307. Provides a list of alternative investment firms.

Institutions and Organizations

Coalition for Environmentally Responsible Economies (CERES), 711 Atlantic Avenue, Boston, MA 02111; (617) 451-0927.

Council on Economic Priorities, 30 Irving Place, New York, NY 10003; (212) 420-1133. CEP's **Institutional Investor Research Service (IIRS)** provides social ratings in eleven categories for all companies in Standard & Poor's 500. CEP's **Corporate Environmental Data Clearinghouse (CEDC)** includes environmental data for a corporation's products and technologies, compliance and pollution record, energy efficiency, waste reduction, and other areas.

Environmental Action, Inc., 1346 Connecticut Avenue, N.W., Suite 731, Washington, D.C. 20036; (202) 833-1845.

Franklin Research and Development Corporation, 711 Atlantic Avenue, Boston, MA 02111; (617) 423-6655.

Interfaith Center on Corporate Responsibility (ICCR), 475 Riverside Drive, Room 566, New York, NY 10115; (212) 870-2936.

Investor Responsibility Research Center (IRRC), Inc., 17555 Massachusetts Ave, N.W., Suite 600, Washington, D.C. 20036; (202) 234-7500.

People for a Socially Responsible University (PSRU), Investment/Divestment Project, Suite 23 Carriage Shops, 233 N. Pleasant St., Amherst, MA 01002; (413) 549-4625.

Williams College, Advisory Committee on Shareholder Responsibilities (ACSR). Committee proceedings available through the Office of the Comptroller, P.O Box 67, Williams College, Williamstown, MA 91267; (413) 597-4411.

CHAPTER 16: ENVIRONMENTAL EDUCATION AND LITERACY

Heightened awareness and concern about the state of the environment among college students has created a new demand for environmental studies programs and curriculum as well as a demand by employers to hire workers trained in a range of environmental professions. In an age of environmental awareness, our institutions of higher education must prepare students to be effective problem-solvers. Education reform is critical to the transition to a sustainable future. The goals of this effort include expanding the number of environmental studies programs and course offerings, promoting environmental literacy by training faculty to incorporate environmental themes into their disciplines, and establishing "Campus and the Biosphere" courses which advance environmental problem-solving through the study of campus resource flows.

As students, you are the market for your school's class offerings. Your assessment survey may be the one way to express your interest in expanding the environmental studies options in your school's curriculum. You may wish to study other schools' offerings before undertaking your assessment.

Assessment Questions

◆ Does your school have an undergraduate environmental studies department? Is it a distinct program or housed in another department? When was the program established? How many courses are offered? Do other departments offer environmentally-related courses? Which departments and how many courses? (See "Research Sources:" A,C,E)

◆ Does your school offer an environmental studies major or minor? How many degrees are conferred in this field each year? Is the number growing or decreasing? By how much? (C,E)

◆ If your school doesn't have a program, are there any plans to establish one? If a program exists, are there plans to expand it? Can you identify interested faculty on your campus who could teach environmentally-oriented courses within existing departments? (C, E)

◆ Does your college or university have graduate environmental degree programs in areas such as environmental engineering, health, hazardous-materials-management, public policy, international environmental policy, environmental law? (C,E)

◆ Do professional schools—medicine, business, architecture, urban planning, public

CAMPUS PROFILE:
Tufts University, Medford, Massachusetts

Educating for Tomorrow

In 1990, Tufts University developed the Environmental Literacy Institute (TELI), which, according to program Dean Anthony Cortese, seeks to promote "environmental literacy" among all the 7,800 students at Tufts' Medford campus by re-educating faculty members about environmental issues in the context of their disciplines. "People who will be making decisions about the environment are graduating from colleges today," says Cortese, "so they need teachers who can educate them and stress the connection between their studies and the environment." As a result of the program, faculty members have revised course curricula to incorporate environmental issues and concerns.

Another goal of the literacy program is to improve the state of the environment within the Tufts community. To this end, the school has launched an ambitious program to minimize the environmental impacts of the university by improving existing programs and initiating new ones. The Tufts Cooperation, Learning and Environmental Awareness Now! (Tufts CLEAN!) is a grant-funded program designed to analyze the energy- and materials-flow throughout the university with the aim of developing cost-effective pollution-prevention strategies.

health, law, engineering, etc.—have concentrations or curriculum offerings in environmental policy or science? (C,E)

◆ Have any environmental studies courses studied the environmental impacts of the campus? Has a "Campus and the Biosphere" type of course been offered? What kinds of research projects have students completed in these courses? Has any of the research resulted in a change of campus policies? (B,C)

◆ Does your school offer student-initiated courses? How have students used this option to design environmental courses? If so, what courses? (C, E)

◆ Have there been any initiatives to provide in-service training for faculty from a broad range of disciplines concerning ways in which they might incorporate environmental themes into their curriculum? (C,D,E)

Research Sources

Ⓐ Your course catalog is a principal resource for a broad review of program and course offerings.

Ⓑ Catalogs from other schools may present possibilities. Contact professors to obtain syllabi for ideas about the content of environmental courses being offered at other campuses.

Ⓒ Your school's existing environmental studies department, school of natural resources, agriculture, and other related departments can help as well. Interview department chairs and other faculty members for additional ideas.

Ⓓ Contact the Tufts Environmental Literacy Program. A goal of the program is to offer assistance to other schools and make the revised curricula available.

Ⓔ The academic affairs office may have information on environmental education and literacy programs.

Recommendations

Meet with curriculum planners and environmental chairpersons. Express your interest in the following:

✔ Establishing introductory environmental studies courses.

✔ Promoting environmental literacy by encouraging instructors in all disciplines to incorporate environmental themes into their courses.

✔ Conducting an "environmental audit" of your campus for course credit. This could range from an undergraduate or graduate course project targeting one or several issue areas to a group or individual master's thesis. Your recommendations can be written up as a formal proposal or faculty/student petition upon completion of your research.

✔ Designing an independent study or student-initiated course in environmental studies.

✔ Establishing an ongoing "Campus and the Biosphere" course, based on the Oberlin College model (request from their Environmental Studies Program).

✔ Consider establishing a major or minor in environmental studies.

RESOURCES

Books, Publications, Articles

Brough, Holly. "Environmental Studies: Is It Academic?" *Worldwatch,* vol. 5, no. 1, January/February 1992.

Capone, Lisa. "Magna Cum Environmentalist: The Environmental Imperative in Higher Education." *E Magazine,* March/April 1991.

"Ecological Literacy: Education for the 21st Century," *Holistic Education Review,* Fall 1989.

Orr, David. "Is Conservation Education an Oxymoron?" *Conservation Biology,* Conservation Education column, vol. 4, no. 2, June 1990.

Orr, David. "The Liberal Arts: The Campus and The Biosphere," *Harvard Education Review,* May 1990.

Orr, David. "Campus and the Biosphere." *Conservation Biology.* Conservation Education column, vol. 3, no. 2, June 1989.

Orr, David. "The Question of Management", *Conservation Biology,* Conservation Education column, vol. 4, no. 1, March 1990.

Institutions & Organizations

Alliance for Environmental Education, 10751 Ambassador Drive, Suite 201, Manassas, VA 22110; (703) 335-1025.

Center for Participant Education, 251 Union, Florida State University, Tallahassee, FL 32302; (904) 644-6576.

Educational Resources Information Center (ERIC), Science, Mathematics, and Environmental Education, Ohio State University, 1200 Chambers Road, Third Floor, Columbus, OH 43212; (614) 422-6717.

Environmental Studies Department, Oberlin College, Oberlin, OH 44074; (216) 775-8312.

North American Association for Environmental Education, 1255 23rd Street N.W., Suite 400, Washington, D.C. 20037; (202) 467-8754.

Student Community Involvement Program, Institute for Environmental Studies, 311 Pittsboro St., CB7410, University of North Carolina, Chapel Hill, NC 27599-7410; (919) 966-3335.

Tufts University, Office of Environmental Programs, Curtis Hall, 474 Boston Avenue, Medford, MA 02155; (617) 381-3486. Ask for the Tufts Environmental Literacy Institute Executive Summary, Fall 1990.

U.S. Department of Education, Environmental Education, Public Affairs Office, Federal Office Building 6, 400 Maryland Ave., S.E., Washington, D.C. 20585; (202) 708-5366.

U.S. Environmental Protection Agency, Environmental Education, 401 M Street S.W., Washington, D.C. 20460; (202) 260-4962.

RESOURCES FOR GRADUATE PROGRAMS IN ENVIRONMENTAL FIELDS

Directory of Post-Secondary Environmental Education, National Environmental/Energy Workforce Assessment, Phase III. National Field Research Center, Inc., Iowa City, May 1979. 230 E. Benton, P.O. Box 287, Iowa City, IA 52240

Freedman, Andrea, *Education for Action: Graduate Studies with a Focus on Social Change,* Institute for Food & Development Policy, 1987. Available from Food First Books, 1885 Mission St., San Francisco, CA 94103; (415) 864-8555. Booklet lists graduate programs in various fields of social change.

U.S. Environmental Protection Agency and U.S. Department of Labor, *Environmental Protection Careers Guidebook,* 1980, Employment and Training Administration, Washington, D.C. 20213. Contains a list of graduate and undergraduate programs divided by state and field.

Annual Publications

Conservation Directory. National Wildlife Federation, 1412 16th St. N.W., Washington, D.C. 20036. List of programs recommended by the NWF in conservation fields.

List of Colleges & Universities with Programs or Interests in Environmental Education. North American Association for Environmental Education, P.O. Box 400, Troy, OH 45373.

Natural Resources & Environmental Management at North American Universities. World Wildlife Federation-US, RARE, and the International Institute for Environment and Development. Contains extensive information about environmental programs at 92 selected schools.

Peterson's Annual Guides to Graduate Study. Very extensive listing and small descriptions of graduate programs broken down by field, including Earth Sciences, Environmental Health, Environmental Design, Environmental Engineering, Environmental Policy and Resource Management, Environmental Sciences, Fish, Game and Wildlife Management, Forestry, Landscape Architecture, Urban Design, etc. Available at most libraries.

Students across the country are deciding that they can no longer tolerate business practices which exploit the environment, and that the ecological consequences of employer actions are important criteria in their career decisions. To this end, more and more students are scrutinizing the environmental implications of their job choices and investigating the records of campus recruiters. More than ever before, students are also exploring occupations in the environmental arena, which have expanded tremendously over the past decade. Emerging jobs in government, the nonprofit sector, and in private industry are creating opportunities for environmental professionals with backgrounds in a variety of disciplines, including engineering, chemistry, biology, journalism, communications, business, law, public health, and planning. Campus career placement centers and academic departments can help students explore environmental careers by organizing job fairs and internships, inviting to campus a broad range of employers working in the field, and by providing extensive resources on environmental careers.

Assessment Questions

◆ What types of companies and organizations recruit at your school? What are their environmental records? Does your career placement center offer interviews with nonprofit environmental organizations? (See "Research Sources:" A,B,C)

◆ Does your career placement office provide resources on environmental jobs? (C,D)

◆ Does your career placement office offer useful information on employers' environmental records, drawing on a variety of independent sources such as the press and nonprofit watchdog organizations, rather than simply gathering company annual reports? (C)

◆ Are staff people in your campus career office knowledgeable about assessing the environmental records of employers? (C)

◆ What kinds of job fairs does your school organize? Is there an environmental job fair? Do law, business, and engineering job fairs include environmental opportunities? (C,D)

◆ Have any companies that recruit at your school adopted the Valdez Principles (see page 82)? (A,B)

◆ Do various departments provide job listings in environmental careers? (E)

CAMPUS PROFILE:
University of Michigan, Ann Arbor

Getting to Work

With over 600 undergraduate and graduate students in the University of Michigan's School of Natural Resources, environmental career planning and placement is more than a full-time job. In fact, it's hundreds of full-time jobs. The decentralized campus has three separate offices that work cooperatively to assist students in finding occupations in the environmental arena. These offices are comprised of Career Planning and Placement, Project Serve, and the School

Research Sources

A "Inside sources" of information about employers include company materials (annual reports, internal newsletters, reports, and other publications) and asking questions during job interviews or meetings with recruiters.

B "Outside sources" of information about employers include advocacy organizations and public interest groups, government regulatory agencies, magazines, journals and newspapers, library databases, and other people who work in your field.

C Counselors at your career placement center can provide information about the types of job opportunities they publicize, including a list of companies that recruit on campus.

D Student and staff organizers can give you information about job fairs.

E Administrative staff and faculty members can tell you if and where environmental career opportunities are listed within their department.

of Natural Resource Placement Services. The overall career placement philosophy stresses the need to develop successful strategies beyond on-campus corporate recruitment.

The university's Career Planning and Placement Office contains a vast library of career descriptions as well as publications containing environmental job listings. Project Serve office is also dedicated to providing students with volunteer work experience with community groups and nonprofit organizations. Volunteer service offers an excellent opportunity to network and create professional relationships that are invaluable to successful job placement. Within the School of Natural Resources, a third office, Placement Services, produces a weekly jobs bulletin containing opportunities found through alumni associations in addition to jobs recruited by the Employer Outreach Publication. This publication, listing majors and student projects and experience, is distributed to over 700 prospective employers.

In 1991, the university held its first Interdisciplinary Environmental Career Conference and is planning to organize this event annually. By combing a variety of student disciplines, from Public Policy to Chemical Engineering, in addition to majors with the School of Natural Resources, the conference has successfully attracted a large number of government agencies, nonprofit and nongovernmental organizations, as well as private corporations.

Recommendations

Here are some ideas for stimulating interest in environmental careers. You can probably think of others. Some possibilities are:

✔ Considering a career for yourself in an environmental field. There are environmental opportunities in a great variety of fields, including waste management, environmental education, urban planning, environmental engineering, air and water quality, and natural resource management. Jobs are available in government, private firms, and the nonprofit sector.

✔ Encouraging your career placement office to invite recruiters that offer jobs in a variety of environmental fields. One contact is the Alternative Chamber of Commerce, an organization created by the Social Venture Network.

✔ Requesting (in writing) that the placement office send a letter to all interviewing companies, inquiring if they have signed or plan to sign the Valdez Principles. Invite a broad coalition of student groups and leaders to sign the letter.

✅ Organizing an Environmental Jobs Fair to expose students to a variety of environmental career opportunities in both the public and private sectors.

✅ Volunteering or interning with a public-interest organization, a government agency, or a private company to gain experience and learn skills in the environmental field.

✅ Organizing a "Graduate Pledge Campaign" in which students are asked to consider the environmental consequences of their job choices (request information from The Graduation Pledge Alliance, below).

✅ Promoting environmental policies and programs on the job. You can promote environmentally responsible business practices at any job you take during the summer months or after college. Encouraging management to start an environmental committee is a good place to start. Many of the recommendations and strategies presented in *Campus Ecology* can be applied to your future workplace.

RESOURCES

Books, Publications, Articles

Anzalone, Joan, ed. *Good Works: A Guide to Careers in Social Change.* Center for the Study of Responsive Law. 1985. Debner Enterprises Corp., 80 Eighth Avenue, New York, NY 10011. A comprehensive directory of jobs in the public interest, publications, networks and training schools. Includes extensive directory.

Berry, Sanford. *Environmental Opportunities.* P.O. Box 969, Stowe, VT 05672; (802) 253-9336.

Bolles, Richard. *What Color is Your Parachute?: A Practical Manual for Job Hunters and Career Changers.* Berkeley: Ten Speed Press, 1984. Updated annually.

The CEIP Fund. *The Complete Guide to Environmental Careers.* Washington, D.C.: Island Press, 1989.

Conservation Directory. National Wildlife Federation, 1400 16th Street, N.W., Dept. GM, Washington, D.C. 20036; 1-800-432-6564. An annual sourcebook of national and international environmental organizations, government agencies, and educational programs focusing on natural resource management.

Earth Work: The magazine for & about people who work to protect the land & the environment. Features Job Scan, listing conservation and natural resources employment opportunities. Published monthly by the Student Conservation Association. Subscription information: SCA, Box 550, Dept. 52G3, Charlestown, NH 03603; (603) 826-4301.

"A Graduation Pledge of Responsibility: Organizing Manual." Graduation Pledge Alliance, P.O. Box 4439, Arcata, CA 95521; (707) 826-7033.

In Business: The Magazine for Environmental Entrepreneuring. Published bimonthly by The Jerome Goldstein Press, Inc., 419 State Ave., Second Floor, Emmaus, PA 18049; (215) 967-4135.

The Job Seeker. Rte. 2, Box 16, Warrens, WI 54666; (608) 378-4290. Newsletter on environmental job openings.

National Directory of Internships. National Society of Internships and Experiential Education, 3509 Haworth Drive, Suite 207, Dept. GM, Raleigh, NC 27609-7229.

Noyes, Dan. *Raising Hell: A Citizen's Guide to the Fine Art of Investigation.* Center for Investigative Reporting. Published by Mother Jones, 1663 Mission St., San Francisco, CA 94103.

Institutions & Organizations

Access, 50 Beacon Street, Boston, MA 02108; (617) 720-JOBS. Provides listings of public interest and nonprofit internships and career opportunities.

Eco-Net, 18 DeBoom Street, Dept. GM, San Francisco, CA 94107; (415) 442-0220. An on-line computer network linking more than 100 electronic bulletin boards on environmental topics, including job listings.

The Environmental Careers Organization, Inc. (Formerly the CEIP Fund), 286 Congress St., Dept. GM, Boston, MA 02210; (617) 426-4375. The largest on-the-job training program for environmental careers. This organization places college students and recent graduates in short-term, paid environmental professions through their Environmental Associate Services program and conducts a variety of seminars through their career service program.

The Graduation Pledge Alliance, Box 4439, Arcata, CA 95521; (707) 826-7033.

Green Corps, 1724 Gilpin St., Denver, CO 80218; (303) 355-1881. Green Corps provides intensive training in grassroots organizing, fundraising, media, and campaign skills, followed by a year of field work.

The Organizing Institute, 1444 I Street, N.W., Suite 701, Washington, D.C. 20005; (202) 408-0700. The Organizing Institute is a premier organizer-training and job placement center for union organizers across the country.

SEAC, P.O. Box 1168, Chapel Hill, NC 27514; (919) 967-4600. SEAC offers many job opportunities and information about available job positions across the country.

Public Interest Research Groups, The Fund for Public Interest Research, 29 Temple Place, Boston, MA 02111; (617) 292-4800. The PIRGs are nonprofit consumer and environmental advocacy organizations with over twenty state offices. They provide numerous employment opportunities for recent college graduates and may recruit at your school.

Social Venture Network, 542 South Dearborn, Chicago, IL 60605; (312) 408-1860. A membership organization for business professionals working for social change.

Student Action Union, P.O. Box 456, New Brunswick, NJ 08903.

Student Conservation Association, 1800 N. Kent St., Suite 1260, Arlington, VA 22209; (703) 524-2441. A nonprofit educational organization providing opportunities for student and adult volunteers to assist in the stewardship and conservation of natural resources.

Student Conservation Association, P.O. Box 550, Dept. 52G3, Charlestown, NH 03603; (603) 826-4301.

Student Pugwash USA, 1638 R Street, N.W., Suite 32, Washington, D.C. 20009; (202) 328-6555; 1-800-969-2784.

Environmental Career Guides

The Complete Guide to Environmental Careers. CEIP Fund. Washington, D.C.: Island Press, 1989.

Conservation Directory. Published annually by the National Wildlife Federation. 1412 16th St. N.W., Washington, D.C. 20036. List of programs recommended by the N.W.F. in conservation fields.

Environmental Protection Careers Guidebook. U.S. EPA and U.S. Dept. of Labor, Employment & Training Administration, Washington, D.C. 20213, 1980. Contains a list of graduate and undergraduate programs divided by state and field.

Exploring Environmental Careers. Stanley Jay Shapiro. New York: Rosen Publishing Group, Inc., 1985.

Opportunities in Environmental Careers. Odom Fanning. 1986. VGM Career Horizons, 4255 W. Touhy Avenue, Lincolnwood, IL 60646.

SECTION IV:
TAKING ACTION

THE KEY TO TRANSFORMING ACADEMIC *research into effective resource management on campus lies in developing and implementing strategies for change. In this section, we'll explore how to create change by communicating across traditional campus boundaries. Students have to learn*

SECTION IV: TAKING ACTION

to work cooperatively with administrators, deans, and regents. Broad-based coalitions must be built between diverse campus organizations. Finally, reaching beyond the campus to forge alliances with community groups is critical to the challenge of achieving environmental justice. Environmental change requires a special combination of vision and pragmatism. For anyone who seeks to create a greener garden in his or her own backyard, campuses are ideal testing grounds for informed activism.

The principles of this book can serve as a platform for moving from study to action. Your research provides the foundation for developing innovative policies and programs to reduce pollution, conserve natural resources, create markets for ecologically sound products, and enhance workplace health and safety. Now you'll need to take steps to ensure that your recommendations become more than just good ideas in a college term paper. Most importantly, implementing environmental policies and programs at your school requires long-term support and commitment from top campus officials. The following ten suggestions will provide guidance on how to create strategies for change:

1) Put it in Writing

Compiling the findings of your research into a concise document is an excellent way to educate the campus community and establish credibility with campus officials. A report can be presented to the administration, released to the press, and shared with other colleges and universities around the country. It can also provide a base of information for developing specific policies and project proposals, such as an environmental investment policy, an energy-efficiency campaign, or a recycling and waste-reduction program.

2) Get a Commitment from the Top

A pledge to campus environmental responsibility from your school's president or chancellor will help create the momentum needed to meet environmental goals. While student and faculty are important for initiating and sustaining operational and educational reforms, they can't replace a commitment from the chief executive and the board of directors. There are numerous ways that your president or chancellor can demonstrate environmental leadership. Signing the Talloires Declaration, a statement of intent to place colleges and universities at the forefront of environmental problem-solving, is a strong first step (see page 108). Creating a full-time position in the administration for an environmental ombudsman, vice-chancellor, or dean of environmental programs will provide the authority for implementing and institutionalizing changes. Establishing an environmental steering committee will provide the basis for policy formation and action. Clearly presenting the benefits of a comprehensive environmental program—such as cost savings, improved community relations and public image, and reduced risk—will also help gain support from campus officials.

3) Creating the Planning Process

If given policy-making authority, an environmental steering committee can help ensure a systematic approach to environmental planning and implementation. A cross-section of the campus community, such as top administrators, physical plant managers, transportation coordinators, food-service personnel, students, and faculty, should all be invited to participate in the committee process. The first task of the committee should be to formulate

CAMPUS PROFILES:

Using Audit Power

Napa Valley Community College, California

In 1990, an Earth Science professor at Napa Valley Community College in California offered students extra credit for conducting the Earth Day Campus Environmental Audit. Upon completion, the group's report kicked off a string of events that lead to reducing waste and conserving natural resources on campus. Students presented their recommendations to the president of the college and the board of trustees. In turn, an environmental committee was established to implement the goals of the audit. Recycling was the first priority, and funds generated from the collected recyclables were used to create a position as student

environmental policies and identify short-term and long-range goals. Policy statements can provide a starting point for implementing ideas and creating departmental procedures. Developing implementation plans and identifying strategies for overcoming institutional and external barriers—such as costs, lack of vendors, product and materials specifications, time constraints, bureaucratic inertia, storage issues, and liability concerns—are also important elements of the planning process. Creating working groups or subcommittees is an effective strategy for formulating policies, procedures, and action plans for specific issues.

4) Fiscal Planning

In a time of shrinking budgets, financial concerns may be the biggest obstacle facing academic institutions in the environmental planing process. Many well-intentioned campus environmental programs won't survive if they lack financial support. However, the notion that all environmental programs cost money is more fiction than reality. In fact, many changes can be extremely cost-effective. Although some programs will require capital

manager for the program. Additionally, four students who conducted the audit were elected officers of the Associated Student Body. According to students at Napa, the audit process also empowered them to help stop the construction of a hotel on 45 acres of local wetlands.

Lansing Community College, Michigan

As part of the requirements for a statistics course at Lansing Community College in Michigan, students used the audit to conduct a waste-characterization study. Working with their professor to develop the methodology, students determined that about one-quarter of the school's waste stream was organic matter—landscape clippings and food waste—that could be composted. Upon completion of the study, students brought their results and recommendations to the attention of the administration, which in turn established a campus composting program.

Oklahoma Baptist University, Shawnee

At Oklahoma Baptist University in Shawnee, students used the Campus Environmental Audit as a mechanism for creating an environmental commissioner position on student government and a task force for reviewing and implement environmental goals. Student efforts on the task force were instrumental in helping the university secure matching funds from the U.S. Department of Energy for a campus energy-efficiency program.

expenditures up front (to buy new glassware for microscale chemistry, to buy low-flow toilets and shower fixtures, or to purchase compact fluorescent bulbs, for example) they provide long-term savings in the form of reduced disposal and utility bills, often with pay-back periods of less than a year. Projecting and documenting cost savings and other economic benefits—such as the creation of student jobs—will help support your recommendations.

Establishing economic incentives is also an important strategy for ensuring program effectiveness. Some ideas include offering a discount on beverages for using reusable mugs, charging individual departments their share of utility bills as an incentive for using energy more efficiently, or rewarding the residence hall with the highest recycling rate. Furthermore, don't underestimate the willingness of students to help pay for environmental programs—such as recycling, for example—through an increase in student fees or paying higher costs for certain items, such as recycled paper notebooks in the student store. Conducting surveys is an effective strategy for gauging the acceptance level of these types of funding mechanisms.

THE TALLOIRES DECLARATION

In October of 1990, twenty-two presidents, rectors, and vice-chancellors of universities from all over the world convened at the Tufts European Center in Talloires, France, to discuss the role of universities and their leaders in environmental management. The group developed and signed the Talloires Declaration as a statement of their commitment to environmental responsibility and challenged their colleagues around the globe to join them. Since 1990, 125 university presidents from 32 countries as well as the Conference of European Rectors have formally endorsed the principles. This is their statement:

WE, THE PRESIDENTS, rectors, and vice chancellors of universities from all regions of the world, are deeply concerned about the unprecedented scale and speed of environmental pollution and degradation, and the depletion of natural resources. Local, regional, and global air and water pollution, accumulation and distribution of toxic wastes, destruction and depletion of forests and soil, depletion of the ozone layer and emissions of greenhouse gases threaten the survival of humans and thousands of other living species, the integrity of the earth and its biodiversity, the security of nations and the heritage of future generations. These environmental changes are caused by inequitable and unsustainable production and consumption patterns of the world.

We believe that urgent actions are needed now to address these fundamental problems and reverse the trends. Stabilization of human population, adoption of industrial and agricultural technologies which minimize resource depletion, pollution and waste and ecological restoration are crucial elements in creating an equitable and sustainable future for all humankind in harmony with nature. Universities have a major role to play in education, research, policy formation, and information exchange to make these goals possible.

5) Work with Campus Administrators

Working cooperatively with the administration throughout your research and advocacy stages is the best way to institutionalize campus environmental programs. Be sensitive to administrative concerns and constraints—such as staff shortages, tight budgets, and time limitations. Listen first, then meet each objection with a solution. Be prepared! If the administration is continually unresponsive to requests for environmental change, however, you may wish to use more assertive tactics. Alternative strategies, such as protests, rallies, guerrilla theater, and boycotts, can be an effective means of initiating reforms by generat-

University heads must provide the leadership and support to mobilize internal and external resources so that their institutions respond to this urgent challenge. We, therefore, agree to take the following actions:

1 Use every opportunity to raise public, government, industry, foundation, and university awareness by publicly addressing the urgent need to move toward an environmentally sustainable future.

2 Encourage all universities to engage in education, research, policy formation, and information exchange on population, environment, and development to move toward a sustainable future.

3 Foster programs to produce expertise in environmental management, economic development, population, and related fields to ensure that all university graduates are environmentally literate and responsible citizens.

4 Foster programs to develop the capability of university faculty to teach environmental literacy and responsibility to all undergraduate, graduate, and professional school students.

5 Set an example of environmental responsibility by establishing programs of resource conservation, recycling, and waste reduction at the universities.

6 Encourage the involvement of government at all levels, foundations and industry in supporting university research, education, policy formation, and information exchange in environmentally sustainable development. Expand work with nongovernmental organizations to assist in finding solutions to environmental problems.

7 Convene deans of appropriate schools and environmental practitioners to develop research, policy, and information exchange programs and curricula for an environmentally sustainable future.

8 Establish partnerships with primary and secondary schools to help develop the capability of their faculty to teach about population, environment, and sustainable development issues.

9 Establish a steering committee and secretariat to continue this momentum and inform and support each other's efforts in carrying out this declaration.

---

ing publicity and getting the administration's attention. Be creative in your approach! For example, you can create a giant garbage sculpture on the steps of the administration building to illustrate the need for campus recycling.

6) Campus Outreach

A diversity of support for your projects will help legitimize them in the eyes of the administration. Student government, fraternities and sororities, student groups, labor unions, the faculty senate, sympathetic campus officials, residence-hall associations,

MEDIA TIPS

Communicating information to the campus community and the public requires knowing how to work with the media. The following guidelines provide some basic tips:

Develop a Media List

✔ Keep a list of planning and assignment editors for radio, television, print media and wire services. List the names of environmental reporters, if any. List the address, phone and fax numbers. You can also include local and national magazines, journals, and newsletters, as well as entertainment media and reporters.

Get to Know Your Campus Newspaper

✔ Getting to know the folks who work on the campus newspaper will increase your chances of getting coverage. Introduce yourself to the appropriate editors and reporters. Set up a meeting with them to inform them about the projects you are working on and the upcoming events which they may be interested in covering in the paper. You may want to discuss with them special features, such as a weekly environmental column or a special campus environmental news series.

Organizing a News Conference

✔ Develop an news "hook" or angle. Before you plan a news conference, make sure your story is newsworthy enough to attract the press in the first place. A press conference is no guarantee that you will get coverage for your event.

✔ Choose an appropriate date, time, and place for your conference. Generally speaking, news conferences held mid-week (Tuesday to Thursday) in the morning (10 a.m.) get the best attendance. This allows editors time to assign a crew and a reporter and gives television time to prepare the story for broadcast. Make sure your news conference does not conflict with other major events on campus or in the community.

✔ Write a Media Advisory. The advisory alerts the press to the "who, what, when, where, and why" specifics of the event. Keep it brief and provide important information such as directions and parking information. Fax the advisory to your media list as early as two days before the event and then again on the morning of the news conference.

✔ Write a Press Release: A press release should be a brief and clear synopsis of what you are trying to communicate. Try to keep it shorter than two pages, double spaced. The press release should read like an article so that it can be printed with few changes. Incorporate a couple of quotes and list a contact person. Fax your press releases on the day of your event.

✔ Prepare a news conference checklist. Identify all the things you'll need for an event, such as parking arrangements, security, press kits, media check-in lists, name tags, photography equipment, audio-visual equipment, podium, refreshments, visuals for the event, restroom access, etc.

✔ Call your media list and pitch your story the day before the event. Ask for the planning editor. Be short and enthusiastic. Get a specific response from them regarding attendance. "We know about it" doesn't mean much. "It's on the boards" most likely means they will be there.

✔ On the day of the event . . . in the morning make calls to secure the RSVP status to all press if you haven't already gotten a "yes." You want to ask for the assignment editor. At the event, have someone assigned to meet the press and have them sign in. You may wish to call your press contacts after the event to see if they need anything. Fax your press release.

Pitching a Story

✔ Write a pitch letter. A news conference is not always the best approach for getting publicity. Another strategy is to pitch your story to a reporter, an editor, or producer. To start, you'll want to write a good pitch letter. The letter should be short (no more than two pages), compelling, and unique. Grab the reader in the first paragraph; remember how many stories come across the editor's desk every day. For television, let them know how good the visuals are.

✔ Contact a selective group of media contacts that you think may be interested in the story. Send them your letter with pertinent background material. Follow up with a phone call. Then start your portfolio of clips when the stories start appearing.

homeowner and tenant associations, and local elected officials can all be approached for support. People and organizations can write letters to campus officials, sign a petition, pass a resolution, write a newspaper editorial, help organize events, or attend an important meeting with administrators. Including a variety of campus constituents and organizations in your organizing efforts also builds ownership and provides a natural vehicle for educating the campus community about environmental issues.

7) Get Students Appointed to Decision-making Positions

Students often have difficulty getting their requests heard because they have little policy-making authority on campus. Getting supportive representatives elected or appointed to decision-making positions is an important democratic strategy for voicing concerns and influencing policies. These positions include seats on student government councils, state-wide student associations, the president's or chancellor's advisory committees, student association board of directors and the board of trustees or regents. If such governing bodies lack student positions, organize a campaign to create them!

8) Monitor Your Progress

Evaluating the progress of campus environmental policies and programs is critical to their long-term success. Monitoring allows you to evaluate program effectiveness and compliance with policies and procedures as well as to measure cost savings and changes in resource flows. Program evaluation will allow your school to identify what works and what doesn't so that appropriate changes can be put in place.

If an environmental audit of your campus has been completed in the past, compare your most recent findings with the original data as a strategy for evaluating your school's progress toward meeting environmental goals. You may want to advocate that your school conduct a comprehensive assessment of the campus environment at least once every five years.

9) Get the Word Out

Whether you are getting the word out to the campus community, the press, or other colleges and universities, communication is one of the most important strategies for creating change. Maintaining a high level of enthusiasm for and participation in environmental programs requires ongoing education and publicity, especially since new students arrive on campus every year. Communication also presents an opportunity to receive recognition for progress, such as an increase in participation levels, quantities of resources conserved, or money saved, for example. All these are important to communicate.

Communication strategies include using the campus newspapers, creating newsletters, organizing publicity events, holding press conferences, and developing orientation and training programs.

10) Use Your Network

Your research and advocacy can be used as an effective model for other academic institutions. In turn, other college and university environmental efforts may help you achieve your goals. Share your information with other campuses, local environmental organizations, and national student environmental groups. Complete the response form in the back of the book and mail a copy to the SEAC National Office so that we can maintain an accurate database on campus environmental initiatives. Also, contact SEAC about the Campus Ecology Project, a program designed by students to further expand the efforts of those who have used this book. The Campus Ecology Project networks students across the country and trains a new generation of people organizing to preserve our environment.

RESOURCES

Organizing Resources

Organizing for Social Change: A Manual for Activists in the 1990s, Midwest Academy, 1991; (312) 645-6010.

Public Interest Research Groups (PIRGs), 29 Temple Place, Boston, MA 02111; (617) 292-4800. The PIRGs have a variety of resources on student organizing and leadership development.

The SEAC Organizing Guide, SEAC, Chapel Hill, NC, 1990.

United States Student Association (USSA), 815 15th Street, N.W., Suite 838, Washington, D.C. 20005; (202) 347-GROW. USSA sponsors the Grassroots Organizing Weekend (GROW), weekend organizing trainings for college students sponsored by campuses across the country.

Media Resources

Bacon's Publicity Checker: Contains over 17,000 publication listings and more than 110,000 editorial contacts for the United States and Canada. Volume 1 lists all magazines organized by industry classifications. Volume 2 lists all daily and weekly newspapers and all multiple publisher groups. 1-800-621-0561.

National Student News Service, 29 Temple Place, Boston, MA 02111; (617) 292-4823.

New Liberation News Service, P.O. Box 41, Cambridge, MA 02139; (617) 253-0399.

Other directories: Media directories exist for major media centers, such as Los Angeles, New York, and Washington. They include the News Media Yellow Book of Washington and New York (Monitor Publishing Co, New York, NY) and the Southern California Media Directory (Publicity Club of Southern California, Sherman Oaks). Also, check your library for state and local media directories.

CHAPTER 21: WORKING FOR ENVIRONMENTAL JUSTICE

While the American environmental movement has predominantly been characterized—and criticized—as being "lily-white," environmental problems transcend racial, cultural, and class boundaries. However, the poor and people of color actually shoulder a disproportionate burden of society's environmental ills. Documenting this link between environmental hazards, poverty, and racism, the 1987 report, "Toxic Waste and Race in the United States" found that three out of five African Americans and Latinos and approximately half of all Native Americans, Asian Americans, and Pacific Islanders live in neighborhoods with toxic waste facilities. In rural America and the inner city, the problems facing poor and minority communities are compounded by such environmental threats as the dumping of radioactive uranium tailings on Native American lands, occupational hazards in industry, lead poisoning, the siting of incinerators and new power plants, and exposure to pesticides in the fields.

Yet a powerful movement is emerging from minority communities across the country for environmental justice—the concept that society's environmental problems should be shared equally. Groups like the "Toxic Avengers of El Puente" (Latino youths of Brooklyn), WHE-ACT (West Harlem Environmental Action), and Oklahoma's Native Americans for a Clean Environment are linking socioeconomic and cultural issues with ecological concerns. Instead of placing the problem somewhere else, organizations like these are fighting for real environmental solutions, such as recycling, energy efficiency, and pollution prevention.

Although campus environmental efforts are still largely populated by white, middle-class students, the cry for environmental justice is being heard on colleges and universities across the country. By building broad coalitions to protect human health and preserve natural resources, students can help create a truly diverse "green" movement both on and off campus.

Ways to Achieve Environmental Justice

✅ Build a broad environmental coalition that includes representatives from special interest and people of color organizations on campus.

CAMPUS PROFILE:
University of Chicago

Building Alliances

In 1989, when the university hospital planned to build an incinerator three times larger than what was needed to burn campus waste, students at the University of Chicago formed a multicultural alliance called SNAP—Student Neighborhood Action Project—and linked up with a local African-American community group called Woodlawn to fight the proposal. The two groups filed a lawsuit and petitioned EPA to reject the application to site the incinerator. They argued that air emissions from the facility would pose a health threat to the campus population and to community residents. Members of the coalition also demonstrated that the project was fiscally unsound and that recycling should be given first priority as an alternative for managing campus waste. After conducting a campus waste-stream analysis, students developed a recycling proposal. All 12 recommendations outlined in the proposal were accepted by the administration within two months, a recycling program was established within a year, and the plan to build an incinerator was defeated.

✔ Develop a common agenda with these organizations. Think of the social, economic, and health implications of each campus environmental project and campaign, as well as the ecological dimensions. Here are some examples:

✔ Pesticides disproportionately effect farm workers, many of whom are people of color or live in low-income communities. Work with community groups, unions, and other groups on campus to get your school to offer organically-grown food and to protect the workers.

✔ Hazardous-waste sites, landfills, and incinerators threaten low-income and communities of color disproportionately. Networking with these groups will strengthen their work to clean up their neighborhoods as well as assist with campus waste-reduction and pollution-prevention efforts.

CAMPUS PROFILE:
New York University, New York City

Communicating Across Campus Boundaries

In 1990, as part of their Earth Day activities, students at New York University conducted an energy audit and advocated energy efficiency on campus as a way to reduce the state's need for multibillion dollar contracts for hydroelectric power from James Bay, Quebec. The controversial project proposes an extensive system of dams and dikes throughout the province that will flood the homelands and threaten and pollute the food sources of the Cree, the Native American tribe indigenous to the region.

✔ The construction of additional power-generating facilities can be prevented with creative conservation measures. Energy efficiency on campus is especially important since construction of power facilities, especially hydroelectric plants, are extremely harmful to the environment and the people living near them.

✔ Organize lectures and forums on environmental justice that emphasize the connections between race, poverty, and the environment.

✔ Encourage the recruitment of women and people of color faculty in environmental studies programs and other departments, and in the administration of your university.

✔ Increase institutional support for people of color, low-income and women students by encouraging additional funding for education and support centers on campus. Work with these constituencies to keep forms of assistance like minority scholarships fully funded.

✔ Involve your campus environmental group in issues affecting local communities. For example, attend hearings on the siting of toxic waste landfills or incinerators planned for low-income or minority communities, talk with immigrant workers in agricultural communities about pesticide issues, and fight to stop the construction of new power plants.

✔ Contact Cool It! and SEAC. Through "Cool It!'s Cultural Diversity" campaign and SEAC's

At NYU and other schools, student activists have formed a diverse coalition with Native American campus groups. NYU campaign organizers have also visited high schools in the Bronx, communicating the message that environmental racism is not just happening in their community, but also in "backyards" around the world.

In the summer of 1990, New York students and community members got a first-hand look at the impacts of the James Bay project when they organized a bike tour throughout the region. The plight of the Cree has also been documented in student film projects at NYU and Cornell. Students have successfully linked with labor unions both on and off campus by highlighting that support for the hydroelectric project sends jobs out of state, at a time when they are desperately needed in New York. Encouraging New York City residents to participate in a half-hour evening "Lights Out" campaign enabled students to outreach to the general public. And students, faculty, and the administration at Spence High School in New York City were so moved by a presentation made by a member of the Cree tribe that they generated 150 letters in opposition to the proposed project and turned off the lights for the rest of the day.
Since Earth Day, students at 150 colleges, universities, and high schools throughout the state have played a key role in organizing an effective campaign to save James Bay, and their petitioning and demonstrating prompted the state's recent decision to cancel its largest contract of $16 billion with Hydro-Quebec.

People of Color Caucus, both of these student networks actively promote environmental justice throughout their program work. Contact them for resources and organizing assistance.

RESOURCES

Books, Publications, Articles

American Council on Education, *Annual Status Report on Minorities in Higher Education.* Washington, D.C.; (202) 939-9300.

"Beyond White Environmentalism: Minorities & the Environment," *Environmental Action,* Jan./Feb. 1990. Also provides a list of local organizations.

Bullard, Robert. *Dumping in Dixie: Race, Class and Environmental Quality.* Westview Press, Boulder, CO 80301; (303) 444-3541.

"The Egg: A Journal of Eco-Justice." Center for Religion, Ethics and Social Policy, Cornell University, Ithaca, NY 14853; (607) 255-4225.

Ruffins, Paul. "Divided We Fall." *New Age Journal,* March/April 1990.

Siting of Hazardous Waste Landfills and Their Correlation with Racial and Economic Status of Surrounding Communities. Washington, D.C.: U.S. General Accounting Office, 1983.

Toxics & Minority Communities. Oakland, CA: Center for Third World Organizing, 1990.

Toxic Waste and Race in the United States. Commission for Racial Justice, United Church of Christ, New York, NY, 1987; (212) 870-2077.

"We Speak For Ourselves: Social Justice, Race and the Environment.," 1990. Panos Institute, 1717 Massachusetts Avenue, N.W., Suite 301, Washington, D.C. 20036; (202) 483-0044.

Institutions & Organizations

Center for Environment, Commerce and Energy, 733 6th St. S.E., Washington, D.C. 20003; (202) 543-3939.

Center for Third World Organizing, 3861 Martin Luther King Jr. Way, Oakland, CA 94609; (510) 654-9601.

Citizens Clearinghouse for Hazardous Waste, P.O. Box 6806, Falls Church, VA 22040; (703) 237-CCHW.

The Community Environmental Health Center at Hunter College, City University of New York, 425 E. 25th St., Box 596, New York, NY 10010; (212) 481-4355.

Cool It! A project of the National Wildlife Federation, 1400 16th St., N.W., Washington, D.C. 20036; (202) 797-6631. Request the packets on cultural diversity.

D.C. Student Coalition against Apartheid and Racism (DC-SCAR), P.O. Box 18291, Washington, D.C. 20005; (202) 483-4593.

Gulf Coast Tenants Association, P.O. Box 56101, New Orleans, LA 70156; (504) 949-4919.

Midwest Academy, 600 W. Fullerton Ave., Chicago, IL 60614.

Migrant Legal Action Center, 2001 S Street, N.W., Suite 310, Washington, D.C. 20009; (202) 462-7744.

National Council of Churches, Eco-Justice Working Group, 474 Riverside Drive, New York, NY 10115; (212) 870-2483.

National Student Campaign Against Hunger & Homelessness, 29 Temple Place, Boston, MA 02111; (617) 292-4823.

National Toxics Campaign, 37 Temple Place, Boston, MA 02111; (617) 482-1477.

Native Americans for a Clean Environment, P.O. Box 1671, Talequah, OK 74465; (918) 458-4322.

Panos Institute, 1717 Massachusetts Avenue, N.W., Suite 301, Washington, D.C. 20036; (202) 483-0044.

Southwest Organizing Project, 211 10th St. N.W., Albuquerque, NM 87102; (505) 247-8832.

Student Environmental Action Coalition (SEAC), People of Color Caucus, Contact the Cultural Diversity Coordinator, P.O. Box 1168, Chapel Hill, NC 27514; (919) 967-4600.

United Farmworks Union of America, AFL-CIO, P.O. Box 62, Keene, CA 93534; (805) 822-5571.

United States Students Association (USSA), 815 15th Street, N.W., Suite 838, Washington, D.C. 20005; (202) 347-GROW. USSA, the nation's largest student organization, hosts Grassroots Organizing Weekends (GROWs) on campuses across the country, leading the way in training a new generation of organizers.

YouthAction, 1830 Connecticut Ave., N.W., Washington, D.C. 20009; (202) 483-1432.

APPENDIX:
ADDITIONAL
RESOURCES

ENVIRONMENTAL PROTECTION AGENCY

Your regional EPA office and the national headquarters are valuable resources on a variety of environmental issues and federal laws. Making use of their libraries, hotlines, databases, and computer bulletin boards will help you with your research.

National Office

EPA Headquarters
401 M Street S.W.
Washington, D.C. 20460
(202) 260-2090

Regional Offices

EPA Region 1
JFK Federal Building
Boston, MA 02203
(617) 566-3715
New England

EPA Region 2
26 Federal Plaza
New York, NY 10278
(212) 264-2515
New Jersey, New York, Puerto Rico,
Virgin Islands

EPA Region 3
841 Chestnut Boulevard
Philadelphia, PA 19107
(215) 597-9800
Delaware, Maryland,
Pennsylvania, Virginia,
Washington, D.C., West Virginia

EPA Region 4
345 Courtland Street N.E.
Atlanta, GA 30365
(404) 347-4727
Alabama, Florida, Georgia,
Kentucky, Mississippi, North Carolina,
South Carolina, Tennessee

EPA Region 5
230 South Dearborn Street
Chicago, IL 60604
(312) 353-2000
Illinois, Indiana, Michigan,
Minnesota, Ohio, Wisconsin

EPA Region 6
1445 Ross Avenue
Dallas, TX 75202
(214) 655-2200
Arkansas, Louisiana, New Mexico,
Oklahoma, Texas

EPA Region 7
726 Minnesota Avenue
Kansas City, KS 66101
(913) 236-2800
Iowa, Kansas, Missouri, Nebraska

EPA Region 8
One Denver Place
999 18th Street
Denver, CO 80202-2413
Colorado, Montana, North Dakota, South
Dakota, Utah, Wyoming

EPA Region 9
215 Fremont Street
San Francisco, CA 94105
(415) 974-8071
Arizona, California, Hawaii, Nevada,
American Samoa, Guam, Trust Territories
of the Pacific

EPA Region 10
1200 Sixth Avenue
Seattle, WA 96101
(206) 442-5810
Alaska, Idaho, Oregon, Washington

COLLEGE AND UNIVERSITY ASSOCIATIONS

The following is a partial listing of college and university associations that may be helpful in conducting and implementing the Campus Environmental Audit. You can find a complete listing in your library in the *Encyclopedia of Associations* under "Higher Education."

American Association for Higher Education
One Dupont Circle, N.W., Suite 600
Washington, D.C. 20036
(202) 293-6440

American Association of Colleges for Teacher Education
One Dupont Circle, N.W., Suite 610
Washington, D.C. 20036
(202) 293-2450

American Association of Community and Junior Colleges
One Dupont Circle, N.W., Suite 410
Washington, D.C. 20036
(202) 728-0200

American Association of Presidents of Independent Colleges and Universities
13800 Biola Avenue
La Mirada, CA 90639
(213) 944-0351

American Association of State Colleges and Universities
One Dupont Circle, N.W., Suite 700
Washington, D.C. 20036
(202) 293-7070

American Association of University Administrators
George Washington University
2121 I Street, N.W.
Washington, D.C. 20052
(202) 994-6503

American Association of University Professors
1012 14th Street, N.W., Suite 500
Washington, D.C. 20005
(202) 737-5900

American Association of University Students
3831 Walnut Street
Philadelphia, PA 19104
(215) 387-3100

American Council on Education
One Dupont Circle, N.W., Suite 801
Washington, D.C. 20036
(202) 939-9310

American Education Research Association
1230 17th Street, N.W.
Washington, D.C. 20036
(202) 223-9485

American Institute of Architecture Students, Inc.
1735 New York Avenue, N.W.
Washington, D.C. 20006
(202) 626-7363

American Society of Landscape Architects
4401 Connecticut Avenue, N.W.
Washington, D.C. 20008
(202) 686-2752

Association for School, College and University Staffing, Inc.
1600 Dodge Avenue, S-330
Evanston, IL 60201-3451
(708) 864-1999

Association of American Colleges
1818 R Street, N.W.
Washington, D.C. 20009
(202) 387-3760

Association of American Universities
One Dupont Circle, N.W., Suite 730
Washington, D.C. 20036
(202) 466-5030

Association of College Unions-International
400 East Seventh Street
Bloomington, IN 47405
(812) 332-8017

Association of Collegiate Schools of Architecture
1735 New York Avenue, N.W.
Washington, D.C. 20006
(202) 785-2324

Association of Collegiate Schools of Planning
Department of City and Regional Planning
University of California
Berkeley, CA 94720
(510) 642-3256

Association of Independent Colleges and Schools
One Dupont Circle, N.W., Suite 350
Washington, D.C. 20036
(202) 659-2460

Association of Physical Plant Administrators of Universities and Colleges
1446 Duke Street
Alexandria, VA 22314
(703) 684-1446

College and University Personnel Association
1233 20th Street, N.W., Suite 503
Washington D.C. 20036
(202) 429-0311

Commission on Independent Colleges and Universities
17 Elk Street; P.O. Box 7289
Albany, NY 12224
(518) 436-4781

Council of Independent Colleges
One Dupont Circle, N.W.
Washington, D.C. 20036
(202) 466-7230

Middle States Association of Colleges and Schools
Commission on Higher Education
3624 Market Street
Philadelphia, PA 19104
(913) 864-5503

National Association of College and University Business Officers
One Dupont Circle, N.W., Suite 500
Washington, D.C. 20036
(202) 861-2500

National Association of College & University Food Services
1450 South Harrison Road, Suite 303
Michigan State University
East Lansing, MI 48824
(517) 332-2494

National Association of Independent Colleges and Universities
122 C Street, N.W., Suite 750
Washington, D.C. 20001
(202) 347-7512

National Education Association
1201 - 16th Street, N.W.
Washington, D.C. 20036
(202) 822-7749

New England Association of Schools and Colleges, Inc.
The Sanborn House
15 High St.
Winchester, MA 01890
(617) 729-6762

North Central Association of Colleges and Schools
Arizona State University
Tempe, AZ 85287-3011
(602) 965-8700

Northwest Association of Schools and Colleges
Commission on Colleges
3700-B University Way, N.E.
Seattle, WA 98015
(206) 543-0195

Society for College and University Planning
2026M School of Education Bldg.
University of Michigan
Ann Arbor, MI 48109-1259
(313) 763-4776

Southern Association of Colleges and Schools
Commission on Colleges
1866 Southern Lane
Decatur, GA 30033-4097
(800) 248-7701

State Higher Education Executive Officers Association
707 - 17th Street, Suite 2700
Denver, CO 80202-3427
(303) 299-3685

Western Association of Schools and Colleges
(Senior Colleges and Universities)
Mills College, P.O. Box 9990
Oakland, CA 94613
(510) 632-5000

Western Association of Schools and Colleges
(Community and Junior Colleges)
P.O. Box 70
Aptos, CA 95001
(408) 688-7575

BOOKS AND PUBLICATIONS

The following resources provide directories of environmental organizations or additional information about environmental law, environmental auditing, and other campus environmental efforts.

Bonine, John E., and Thomas O. McGarity, *The Law of Environmental Protection,* St. Paul, Minn: West Publishing Co., 1984.

Brink, Tamra, et al. *In Our Backyard: Environmental Issues at UCLA: Proposals for Change and the Institution's Potential as a Model.* Graduate School of Architecture and Urban Planning, UCLA, June 1989.

Callenbach, Ernest, et al. *The Elmwood Guide to Eco-Auditing and Ecologically Conscious Managment,* Global File, Report No. 5, The Elmwood Institute, P.O. Box 5765, Berkeley, CA 94705; 1990. (510) 845-4595.

Eagan, David, and David Orr, eds. *Campus and the Biosphere: Emerging Opportunities for College and University Environmental Leadership,* New Directions for Higher Education, Jossey-Bass Inc., 350 Sansome Street, San Francisco, CA 94104; 1992. (415) 433-1740.

Findley, Roger W., and Daniel Farber. *Environmental Law.* Nutshell Series. St. Paul, Minn: West Publishing Co., 1988.

Pearlman, Nancy Sue, ed. *1991 Directory of Environmental Organizations.* Educational Communications. P.O. Box 35473. Los Angeles, CA 90035. (213) 559-9160.

SEAC. *The Student Environmental Action Guide: 25 Simple Things We Can Do.* Berkeley: The EarthWorks Press, 1991.

Selected Environmental Law Statutes: 1989–90 Educational Edition. St. Paul, Minn: West Publishing Co., 1989.

Seredich, John, ed. *Your Resource Guide to Environmental Organizations.* Smiling Dolphins Press, 4 Segura, Irvine, CA, 92715; 1991. (714) 733-1065.

U.S. EPA, *Access EPA,* National Technical Information Service, Washington, D.C., 1992; (703) 487-4650. This publication consolidates many EPA directories, hotlines, and databases.

April Smith grew up in Bennington, Vermont and Atlanta, Georgia and graduated from the University of Vermont in 1986 with a B.A. in History. After working with the Public Interest Research Groups in California for two years, April returned to school and received a Master's degree from UCLA in Urban Planning. As a graduate student, April was a member of the research team that wrote the award-winning master's thesis, *In Our Backyard: Environmental Issues at UCLA—Proposals for Change and the Institution's Potential as a Model.* She is the creator of the *Campus Environmental Audit,* which served as the centerpiece for Earth Day 1990s national student campaign and the basis for *Campus Ecology.* April currently works in Los Angeles as an environmental planner in the entertainment industry, in government, and with academic institutions, nonprofit organizations, and private businesses. She is the co-founder of E^2 Environmental Enterprises, a consulting firm specializing in environmental building design and corporate environmental management.

SEAC
MEMBERSHIP FORM

The Student Environmental Action Environmental Coalition is young people like yourself—leaders of tomorrow working to build a global student environmental movement today!

☐ **YES! I would like to be a SEAC individual member.** Here is $15 to receive *Threshold* and other member benefits for one year.

☐ **YES! Our group would like to be a SEAC group member.** Here is $35 so we can join the network and receive *Threshold* and other important mailings for one year.

Name _____

Group name _____

School name _____

School address _____

Permanent address _____

Phone (home) _____ (work) _____ Fax _____

We urge all students to join as individuals even if their group is already a member (group membership is now required to be a part of the SEAC network). If you are unable to pay $15, call or write and we can work out a reduction/waiver of the fee. Return to:

SEAC, P.O. Box 1168, Chapel Hill, NC 27514; (919) 967-4600

History

In the Spring of 1989, the environmental group at the University of North Carolina in Chapel Hill began networking with student environmentalists across the country in the hope of building a national student and youth environmental movement. Out of these efforts, the Student Environmental Action Coalition (SEAC) was born. In just three years SEAC has expanded into the largest student-run organization in the U.S. with members at over 2,200 universities, colleges, and high schools in all fifty states.

Mission

SEAC is a grassroots network working for environmental protection and social justice. SEAC is guided solely by the needs and efforts of students and youth, and all of SEAC's programs are designed and coordinated by young people around the country. SEAC is committed to creating a diverse student environmental movement that unites people in the fight for environmental justice.

Resources and Benefits

Becoming active in the Student Environmental Action Coalition gives you direct access to SEAC's wide range of student-oriented programs:

◆ SEAC publishes *Threshold,* a monthly newsletter that has been called the "Bible of student action." *Threshold* keeps young people up-to-date on the most exciting environmental campaigns around the country and gives advice on issues ranging from recycling to fighting environmental racism.

◆ SEAC sponsors over 70 student gatherings each year in different parts of the country. These conferences are great places to meet other young people fighting to save the planet, to receive training, information—and to have fun!

◆ SEAC maintains a network of Field Organizers—experienced student leaders who travel to campuses to hold workshops, sponsor trainings, and support work on campus audits and other campaigns—as well as strengthen the national network of young environmentalists.

◆ SEAC is part of A SEED, (Action in Solidarity, Equality in Environment and Development), the largest grassroots environmental youth network in the world. Through A SEED, SEAC members communicate and cooperate with young people working for environmental justice around the globe, from Malaysia to Kenya to Uruguay.

◆ SEAC's structure includes 17 regional networks and 50 state networks, each with student coordinating groups that keep members in touch with each other and with student campaigns.

◆ SEAC members have made a positive impact all over the nation, from creating over 700 recycling programs, to holding the largest student environmental conference ever (7,600 young people!), to serving on the official U.S. delegation to the U.N. Earth Summit.

◆ SEAC is networking hundreds of student environmental groups across the country through the Campus Ecology project, which supports students using *Campus Ecology* in their pursuit for environmental change.

SEAC is young people like yourself—leaders of tomorrow working to build a global student environmental movement today!

For more information: Call (919) 967-4600 or fill out the membership form on the facing page and mail it to us.

ABOUT COOL IT!

A PROJECT OF
NATIONAL WILDLIFE
FEDERATION

Earth Day Every Day ®

History

In early 1989, the National Wildlife Federation issued a challenge to college faculty, students, and administrators from diverse cultural backgrounds to support Earth Day 1990 by starting environmental programs on campuses and in their communities. By Summer 1989, Cool It! was established as the Campus Outreach Division of the National Wildlife Federation. Two and a half years and approximately 500 projects later, Cool It!'s commitment to environmental action and leadership training has well outlasted Earth Day 1990.

Mission

Cool It!'s mission is to motivate college leaders to establish national campus and community models of environmentally sound practices through a culturally-inclusive process. Cool It! provides factual resources, organizing tools and one-on-one consultation to students, faculty, and administrators as part of N.W.F.'s effort to promote responsible stewardship.

Resources and Benefits

Cool It! offers a broad array of resources and benefits to members. Campus leaders need only fill out a short questionnaire and enclose a requested $10.00 membership fee. Resources and benefits of membership include:

◆ **Issue Packets:** We currently have packets on Composting, Avoiding Energy-Intensive Foods, Procurement, Recycling, Fundraising, Tree Planting, Energy Efficiency, Cultural Diversity, and Organizing.
◆ **One-on-One Consultations and Site Visits:** Our regional coordinators work with students over the phone from week to week. And each Cool It! organizer makes an effort to conduct on-site workshops tailored to the needs of the campuses s/he is serving.
◆ **Cool It! Connection Letter:** A free publication to our newsletter alerts students to the latest information on pertinent issues.
◆ **Job Bank:** We assist students in their job searches by providing a list of environmental organizations' addresses and phone numbers, acting as a clearinghouse for job information.
◆ **Speakers Bureau:** You name your topic and we search for a speaker for you.
◆ **The Cool It! Directory:** Provides detailed documentation for campus models.
◆ **Econet:** On-line computer environmental information system.

For more information: Cool It!, National Wildlife Federation, 1400 16th Street, N.W., Washington, D.C. 20036; (202) 797-5435.

Keeping the Earth Day spirit alive

History

In the aftermath of Earth Day 1990, many local organizers expressed interest in working with a permanent national Earth Day organization. Earth Day Resources was started in early 1991 to provide support to these organizers and to ensure that there would be a grassroots group to keep the spirit of Earth Day alive. The group is also committed to promoting environmental programs throughout the year. Earth Day Resources is chaired by Christina Desser, who was the Executive Director of Earth Day 1990.

Mission

Earth Day Resources' mission is three-part:

1) To maintain and support a network of organizers of Earth Day events who are working to promote environmental responsibility at the local level.
2) To act as a national resource center and clearinghouse for year-round environmental activities. Earth Day Resources provides support to community organizers, educators, and college activists in the form of materials and organizing assistance.
3) To ensure that Earth Day maintains a grassroots nature. Earth Day Resources works to promote Earth Day as it was originally intended, as a demonstration of concern and a chance to find real solutions to environmental destruction.

Materials and Services

Earth Day Resources offers the following materials to promote environmental action in several areas:

◆ **Lesson Plans:** These curriculum booklets (grades K-6 and 7-12) each cover 4 topics (water, energy, recycling, and household hazardous materials) and help the teacher to design the lesson with background materials, tips on leading a discussion, and follow-up exercises.
◆ **Projects for the Planet:** These kits are designed to help organize workplace environmental programs. Each includes a "how-to" manual and two posters that can be photocopied to help promote the program:
Project #1: How to Organize a Workplace Recycling Program
Project #2: How to Organize a Workplace Ridesharing Program
◆ **Fact Sheets:** Basic information on the major environmental issues facing the planet and our communities.
◆ **Planning and Running Your Environmental Event:** Prepared by Stanford's Haas Center for Public Service, this booklet is a thorough guide to organizing an Earth Day or any environmental event, including publicity, fundraising, and getting volunteers.
◆ **Workplace Environmental Audit:** A brief guide to understanding environmental problems and solutions at work.

Earth Day Resources also provides media support each Earth Day and sponsors an annual Earth Day petition on a critical environmental issue, to encourage Earth Day participants to have a voice in policy decisions.

For more information: Earth Day Resources, 116 New Montgomery St. Suite 530, San Francisco, CA 94105; (800) 727-8619 or (415) 495-5987.

CAMPUS ENVIRONMENTAL AUDIT RESPONSE FORM

The following Campus Environmental Audit Response Form is a modification of the assessment questions presented throughout the book and is intended to provide information to SEAC about campus environmental policies and practices so that they may share this information with students across the country. You may also find it to be a useful questionnaire for collecting information from campus officials.

We urge you to fill out this response form! This information will help us maintain an up-to-date database that can be shared with other colleges and used to track the progress of campus environmental efforts across the country.

Instructions:

Tear out this form, collect the information with the assistance of *Campus Ecology*, and send it to SEAC, P.O. Box 1168, Chapel Hill, NC 27514-1168. If you need more space to answer the following questions, attach additional sheets to the response form or copy the questions onto a computer disk.

GENERAL CAMPUS INFORMATION

1) Name of School _____

2) Address _____

   State _____ Zip code _____ Phone number _____

3) Is your campus public or private?   PUBLIC   PRIVATE

4) Is your campus located in a rural, suburban, or urban area?   RURAL   SUBURBAN   URBAN

5) Number of students on campus:
   Undergrad _____ Grad _____ Part-time _____ Full-time _____

6) Number of employees on campus:
   faculty _____ non-student staff (non-faculty) _____ student employees _____

7) Total campus population _____

8) Total acreage _____

9) Building space (total square feet) _____

10) Are there any environmental organizations or committees on campus?   YES   NO

11) Does your school have an environmental ombudsman or vice-chancellor?   YES   NO

    Please name and describe. _____

    _____

    _____

12) What are the names and affiliations of the members of the Board of Trustees or Regents?

    _____

    _____

    _____

    _____

WASTES AND HAZARDS

Solid Waste

13) How much total solid waste does your campus generate annually? _____

14) Have any waste-composition-studies been conducted?   YES   NO

    If so, please provide information on the composition of campus solid waste. _____

    _____

    _____

15) For the past academic year, how much solid waste was:

    Landfilled? _____ incinerated? _____ recycled? _____ composted & mulched? _____
    (Information should be given by volume, in cubic yards, or by weight, in pounds or tons.)

16) What were the costs of solid-waste disposal for the last academic year? _____

17) How have they changed over the past five years? _____

    _____

18) Does your campus have a recycling program?   YES   NO

    When was the program started? Who operates the program? _____

    _____

    _____

19) What is the budget for the program? How is it funded? Indicate revenues from recyclables sold.

    _____

    _____

20) If your school has a recycling program, how many tons of each material were recycled during
    the last academic year?

    newsprint _____ glass _____ white ledger/ computer paper _____ mixed/ colored paper _____

    aluminum _____ other metals _____ cardboard _____ plastic _____

21) Please describe any programs your campus has implemented to promote source reduction (such as a reusable mug program, double-sided copying policies, switching from disposable to washable dishes, etc).

_____

_____

_____

Hazardous Substances

22) What kind of hazardous waste does your school generate and what are the sources? _____

_____

_____

23) How much hazardous waste does your campus generate annually? _____

24) How has this figure changed over the past five years? _____

_____

25) How is this waste disposed of? _____

_____

_____

26) How much is recycled? _____ incinerated? _____ landfilled? _____

27) What were the total hazardous-waste disposal costs for the last academic year? _____

28) How have these costs changed over the past five years? _____

_____

29) What is being done on campus to minimize the quantity of hazardous substances used and waste generated?

_____

_____

_____

30) Have microscale chemistry techniques/surplus chemical exchange programs been initiated?
YES   NO

If so, please describe, including date of implementation and cost-savings to date. _____

_____

_____

31) If the Chemistry Department has implemented a microscale laboratory program, how many courses use microscale techniques and how many students does this include?

_____

_____

32) Roughly, what portion of the chemistry program does this represent?

ALL   MOST   MORE THAN HALF   LESS THAN HALF   ONLY A SMALL FRACTION

33) Does your school have a system for tracking and inventorying hazardous chemicals bought and used?   YES   NO

If so, please describe. _____

_____

_____

Radioactive Waste

34) What departments and activities on your campus generate radioactive waste and use radioactive substances?

_____

_____

_____

35) What are the quantities of radioactive substances used and wastes generated on campus annually?

_____

36) How has this figure changed over the past five years? _____

_____

37) How and where is radioactive waste disposed of and where does it go? _____

_____

38) How much is landfilled? _____

  incinerated? _____

39) Does your school have an on-site radioactive-waste incinerator?   YES   NO

40) What were the total radioactive-waste disposal costs for the last academic year? _____

41) How have these costs changed over the past five years? _____

  _____

42) Has your campus initiated a radioactive-waste reduction program?   YES   NO

  If so, please describe, including implementation date and cost-savings to date. _____

  _____

  _____

Medical Waste

43) What campus facilities generate medical waste? _____

  _____

  _____

44) How much medical waste does your campus generate annually? _____

45) How has this figure changed over the past five years? _____

  _____

46) How is medical waste disposed of and where does it go? _____

  _____

  _____

47) How much is landfilled? _____ incinerated? _____

48) What are the annual medical-waste disposal costs? _____

49) How have these costs changed over the past five years? _____

  _____

50) Approximately, what percentage of the medical waste is recyclable? _____

51) Have any efforts been implemented to separate recyclable materials from non-recyclable medical waste? YES NO

If so, please describe. _____

_____

_____

52) Has your campus initiated a medical-waste reduction program? YES NO

53) If so, please describe, including implementation date and cost-savings to date. _____

_____

_____

Wastewater and Storm Runoff

54) How much wastewater (sewage) does your campus generate annually? _____

55) Where is campus wastewater treated? _____

_____

_____

56) Where is this treated wastewater discharged? _____

_____

57) What percentage of the capacity of your community's wastewater treatment facility is used to process campus-generated wastewater?

_____

58) What kind of treatment does it receive? _____

_____

_____

59) What costs, if any, are associated with treating campus wastewater? _____

60) Has your campus initiated any programs to reduce wastewater volume and/or toxicity?
YES NO

If so, please describe. _____

_____

61) Does your school use any reclaimed water in its facilities or on landscaping?   YES   NO

If so, how much? _____

62) What is the source of this water? _____

_____

63) Is campus storm water runoff treated, or does it flow directly to a body of water?
TREATED   UNTREATED

Pest Control

64) What pesticides are used on campus and in what quantities? _____

_____

_____

65) How has this quantity changed over the past five years? _____

_____

66) What kind of pests are they controlling and where? (Specify indoor and outdoor applications.)

_____

_____

67) What are the total costs for contracts and purchases related to pest control? _____

68) How have these costs changed over the past five years? _____

_____

69) Has your campus initiated an Integrated Pest Management program?   YES   NO

70) If so, please describe, including implementation date and cost-savings to date. _____

_____

_____

Air Quality

71) Are federal air-quality standards exceeded in your area?   YES   NO

If so, which ones? _____

72) How often? _____

73) What stationary sources on campus emit toxic air pollutants? _____

_____

_____

74) How much air pollution does your campus generate annually? _____

75) What are the most common pollutants? (Distinguish between stationary and mobile sources.)

_____

_____

76) Has your campus initiated programs to reduce air pollutants from stationary sources? YES   NO

If so, please describe, including implementation date. _____

_____

_____

77) What kinds of ozone-depleting compounds are used on campus and what are the sources?

_____

_____

78) Are any programs in place to recapture ozone-depleting compounds or find safer alternatives?
YES   NO

The Workplace Environment

79) What kinds of occupational environmental hazards exist on campus? _____

_____

80) Does your campus have a history of worker/student health and safety complaints or problems?
YES   NO

81) What kind of training is required for faculty and students exposed to environmental hazards in departments such as chemistry, medicine, or art?

_____

_____

82) Are any courses or seminars conducted on art safety for students?   YES   NO

for faculty?   YES   NO

83) Is there a non-smoking policy in effect on campus?   YES   NO

84) Have workers or students complained of health problems that may be due to poor indoor air quality?
YES   NO

85) What are the sources of poor indoor air quality on campus? _____

_____

86) Have campus officials investigated indoor air-pollution problems, including radon?   YES   NO

87) What is the status of asbestos abatement in campus facilities? _____

_____

_____

88) Is water quality a concern in your community?   YES   NO

89) What is the quality of drinking water on campus? Has it been tested for contaminants, such as lead?

_____

_____

90) Has your school developed policies and procedures regarding the use of Video Display Terminals (VDTs)?   YES   NO

91) Is your school located near a power-generating station?   YES   NO

92) Is there any cause for concern on campus regarding electromagnetic fields (EMFs)?   YES   NO

RESOURCES AND INFRASTRUCTURE

Water

93) Where does your campus water supply originate? _____

_____

94) Are water-supply issues a concern in your region?   YES   NO

95) How many total gallons of water did your school consume last year? _____

96) How many gallons did your school consume:
   per square foot of building space? _____ per capita? _____

97) How has this figure changed over the past five years? _____

_____

98) What percentage of water is used indoors versus outdoors? _____

99) What were the water utility costs for the university for the last academic year? _____

100) How have these costs changed over the past five years and why? _____

_____

101) Does the campus have a water conservation program?   YES   NO

   If so, what measures are included?

   ❑ Performing waste audits
   ❑ Leak detection and repair
   ❑ Low-flow showerheads in new restrooms   ❑ in existing restrooms
   ❑ Low-water landscaping (xeriscaping) in new landscaped areas   ❑ in existing areas
   ❑ Efficient irrigation system (circle whichever ones apply):
        drip   automatic timers   automatic sensors
   ❑ Use of reclaimed water
   ❑ Others: _____

   _____

102) Are there any estimates of water savings from such programs?   YES   NO

   If so, how much? _____

Energy

103) What are the sources of energy for the electric utility serving your school? _____

_____

_____

_____

104) How much energy did campus buildings and grounds consume in the last academic year, and what were the costs associated with each type of fuel? (Indicate totals and square-foot consumption.)

Electricity: total use (KWHs) _____ total cost _____ cost/KWH _____

105) Natural gas: total consumption (BTUs) _____ total cost _____ cost/BTU _____

106) Fuel oil: total use (gals. or BTUs) _____ total cost _____ cost/BTU or gals. _____

107) How has campus energy use changed over the past five years? _____

_____

Why? _____

_____

108) Does your campus have an energy-efficiency program?   YES   NO

Which of the following programs does it include?

❑ Regular energy audits of campus buildings
❑ Replacement of incandescent lighting with more efficient lighting systems
❑ Computerized energy-management system
❑ Energy-conservation-awareness program for staff and/or on-campus residents
❑ Solar water heating
❑ Passive solar building design
❑ On-campus cogeneration
❑ Energy-efficient windows
❑ Energy-efficient appliances
❑ Other: _____

_____

_____

Food

109) Who operates your campus food service? _____

_____

110) Does your campus administration manage the operation or do they contract for services?
MANAGE   CONTRACT

111) Does your school have a food-service committee?   YES   NO

112) Who sits on the committee? _____

_____

113) To what extent does food services purchase from regional growers and food processors?

_____

_____

114) Does your food service operation offer vegetarian meals on a regular basis?   YES   NO

115) Does your school purchase certified organically-grown produce or meat and dairy products?
YES   NO

116) Have any surveys been conducted to estimate the demand for vegetarian food on campus?
YES   NO

If yes, what were the results? _____

_____

117) Have food services discontinued the purchase of any food products for environmental reasons?
YES   NO

118) Have any programs or events taken place on campus highlighting the connection between diet and the environment?   YES   NO

Procurement Policies

119) How many reams or tons of high-grade writing paper and copy paper does your campus purchase annually?

_____

120) What is the associated cost? _____

121) Does the campus, one or more departments, or student association have a policy of preferentially buying products—such as paper products, building materials, oil, and tires—made from recycled materials instead of virgin materials?   YES   NO

122) Please specify the material and the quantity of the purchase. What is the cost difference between these products and their virgin equivalents?

_____

_____

_____

123) Do campus food services use primarily disposable plastic and paper products, washable dishes, or a combination of both?   DISPOSABLE   WASHABLE   COMBINATION

124) Are any tropical hardwoods used in the construction of new buildings or other campus fixtures, such as landscape trellises?   YES   NO

125) Does your school purchase any tropical hardwood furniture?   YES   NO

126) What kind of wood is used and where does it come from? _____

_____

127) What programs and policies have been established on your campus to promote the use of ecologically sound products (such as organic produce, compact fluorescent light bulbs, non-toxic cleaning products, etc)?

_____

_____

_____

Transportation

128) How do people get to campus every day? If possible, provide estimates of the percentage by each mode:

single-occupancy vehicle _____   carpool _____
vanpool _____   transit _____
bicycle _____   walking _____
other (please specify) _____

129) How many vehicles travel to campus daily? _____

130) How far do people commute on average? _____

131) What percentage of your campus area is devoted to roads and parking lots? _____

132) Does the campus have a program to promote ride-sharing? (e.g. carpool, matching services, preferential parking, reduced parking rates, subsidized vanpools, etc.)   YES   NO

    If so, please describe. _____

    _____

133) Are transit passes subsidized for students, staff, or faculty?   YES   NO

    If so, please describe. _____

    _____

134) Does the campus have enough parking to meet the demand?   YES   NO

135) Is it campus policy to provide parking to meet demand?   YES   NO

136) Are students, staff, and faculty charged for parking?   YES   NO   How much? _____

137) Are any campus fleet vehicles operated using alternative fuels (e.g. propane, methanol, natural gas)?   YES   NO

138) Does your school own or lease any electric vehicles?   YES   NO

    If so, please describe. _____

    _____

Campus Design and Growth

139) When was your school established? _____

140) How much land is now owned or operated by your school and what is its geographical dispersement?

    _____

    _____

141) What are your school's current plans for expansion or renovation? _____

    _____

142) Does your school have a long-range development plan describing existing and future land uses for campus?   YES   NO

   If so, does the document contain environmental criteria?   YES   NO

143) Does your school have an ongoing planning committee?   YES   NO

144) Who sits on the committee? Please describe. _____

   _____

   _____

145) Is the campus exempt from any local (city or county) land-use planning and zoning laws?
   YES   NO

146) Are there, or have there been, any land-use conflicts between the campus and the surrounding community?   YES   NO

   If so, how were they dealt with? Please describe. _____

   _____

   _____

147) Does the campus own any land that will be developed for private, noneducational facilities (e.g. office park, shopping center, non-staff or student housing, etc.)?   YES   NO

   If so, please describe. _____

   _____

148) Are there any examples on campus of environmental building design?   YES   NO

   If so, please describe. _____

   _____

   _____

149) Does your school offer cooperative housing?   YES   NO

150) How have environmental principles been incorporated into these housing arrangements?

   _____

   _____

THE BUSINESS OF EDUCATION

Research Activities

151) What are the largest research projects at your school, in terms of funding? _____

_____

_____

152) What are the titles of the projects, affiliated campus departments, and the levels of funding?

_____

_____

153) Name the research projects that deal with the subject of toxics or hazardous waste, including research on the generation, use, and disposal of such materials, as well as occupational and environmental hazards involved.

_____

_____

154) Are there any projects to research pollution prevention or hazardous-waste-reduction?

YES   NO

155) If so, please provide a short description of the projects. _____

_____

_____

156) What are the funding levels for these projects? _____

157) Who are the project sponsors? _____

_____

_____

158) Who are the participating faculty and departments? _____

_____

_____

159) Name the research projects that deal with the subject of solid waste. _____

_____

_____

160) What are the funding levels for these projects? _____

161) Who are the sponsors? _____

_____

_____

162) Who are the participating faculty and departments? _____

_____

_____

163) Please provide a short description of the projects. _____

_____

_____

164) Are there any projects to research solid-waste-reduction and recycling?   YES   NO

165) Describe any other environmentally-related research efforts (e.g. alternative energy, organic agriculture, xeriscaping, etc) Please provide information on funding, sponsors, and participating faculty and departments.

_____

_____

166) Does your campus have Department of Defense contracts?   YES   NO

What is the nature of this research? _____

_____

_____

167) Does your campus have Department of Energy contracts?   YES   NO

What is the nature of this research? _____

_____

_____

Investment Policies

168) What is the total amount of your school or school system's investment holdings? _____

169) Does your school rank among the largest investors in any particular company?   YES   NO

170) What are the names of the top ten companies (in value of shares) that your school or school system invests in? Assign number of shares, their "book" (original) value, and their current value.

_____

_____

_____

171) Does your school invest in corporations with poor environmental records?   YES   NO

If so, which companies? What are the number of shares held and their value? _____

_____

_____

172) Does your school have an Environmentally or Socially Responsible Investment policy?
YES   NO

Please describe. _____

_____

_____

173) Does your school screen investments for any social or environmental reasons?   YES   NO

Please describe. _____

_____

_____

174) Has your school divested from any companies for social or environmental reasons?   YES   NO

Please describe. _____

_____

175) Has your school used its proxy votes to try to change any social or environmental policies and practices of any companies?   YES   NO

Please describe. _____

_____

176) Has your school actively invested in environmentally responsible companies?   YES   NO

Please describe. _____

_____

_____

177) Does your Board of Trustees/Regents have a student representative?   YES   NO

Business Ties

178) Does your school have a policy that makes the environmental practices of a company a criterion in selecting campus suppliers and vendors?   YES   NO

179) Has your school chosen to stop doing business with a particular company because of environmental reasons?   YES   NO

If so, which ones and why? _____

_____

_____

180) What companies have the top ten largest contracts with your school? _____

_____

_____

181) What are their environmental records? Do these companies have environmental policy statements?

_____

_____

182) Does your school have an Environmental or Social Responsibility Policy which enables the campus community to formally scrutinize the companies your school does business with?
YES   NO

Environmental Education and Literacy

What are the types of environmental studies programs offered at your school? Please provide the following information for each program:

183) Name of program _____

184) Department _____

185) School _____

186) Undergraduate degree offered ?   YES   NO   Graduate degree offered?   YES   NO

187) Annual budget _____

188) Year program was created _____

189) Enrollment _____

190) Number of FTEs _____

191) Program Chair:
Name _____

Address _____ Phone _____

192) Please list specific course offered in the following subjects:

❑   Toxics/hazardous waste _____

❑   Solid waste _____

❑   Energy efficiency and management _____

❑   Water conservation _____

❑   Air quality _____

❑   Water _____

❑   Environmental economics _____

❑ Environmental law _____

❑ Environmental business _____

❑ Forest management _____

❑ Other _____

193) Which, if any, environmental courses have included the study and research of campus environmental issues in the curriculum?

_____

_____

194) Describe any student research projects (include dates) that have resulted from these courses and any campus environmental programs that were subsequently implemented.

_____

_____

195) Do any environmental literacy programs exist on campus that train faculty to incorporate environmental themes into their curricula?   YES   NO

Job Placement and Environmental Careers

196) Does your Career Placement Center (CPC) or other entity organize job fairs focused on environmental careers?   YES   NO

197) Does your CPC provide information on environmentally-oriented jobs?   YES   NO

198) Does your center place access fees on a sliding scale so that small environmental firms and non-profit organizations can afford to recruit university graduates?   YES   NO

199) Have students organized any boycotts against particular companies recruiting on campus for environmental reasons?   YES   NO

If so, which companies and why? _____

_____

200) Do you know of any efforts on campus to get students to sign a pledge encouraging them to work for socially responsible companies upon graduation?   YES   NO

RESPONDENT INFORMATION

Please list below the student(s) responsible for completing the Campus Environmental Audit Response Form:

201) Name _____

Year in school _____

Campus address _____

Campus phone number _____

Permanent address _____

Permanent phone number _____

202) Name _____

Year in school _____

Campus address _____

Campus phone number _____

Permanent address _____

Permanent phone number _____

Date of completion _____

WE WANT TO HEAR FROM YOU!

Please let us know how *Campus Ecology* was used at your school and how it can be improved for future editions:

_____

_____

_____

_____